Common Core Subject Test

Mathematics Grade 4

Student Practice Workbook

+ Two Full-Length Common Core Math Tests

SCAN ME

Math Notion

www.MathNotion.com

Common Core Subject Test Mathematics Grade 4

Published in the United State of America By

The Math Notion

Web: WWW.MathNotion.com

Email: info@Mathnotion.com

SCAN ME

ISBN: 978-1-63620-092-7

The Math Notion

Michael Smith has been a math instructor for over a decade now. He launched the Math Notion. Since 2006, we have devoted our time to both teaching and developing exceptional math learning materials. As a test prep company, we have worked with thousands of students. We have used the feedback of our students to develop a unique study program that can be used by students to drastically improve their math scores fast and effectively. We have more than a thousand Math learning books including:

– SAT Math Prep

– ACT Math Prep

– SSAT/ISEE Math Prep

–Mathematics Prep Grade 3 to 8

– Common Core Math Prep

–many Math Education Workbooks, Study Guides, Practice and Exercise Books

As an experienced Math test preparation company, we have helped many students raise their standardized test scores—and attend the colleges of their dreams: We tutor online and in person, we teach students in large groups, and we provide training materials and textbooks through our website and through Amazon.

You can contact us via email at:

info@Mathnotion.com

Get the Targeted Practice You Need to Ace the Common Core Math Test!

Common Core Subject Test Mathematics Grade 4 includes easy-to-follow instructions, helpful examples, and plenty of math practice problems to assist students to master each concept, brush up their problem-solving skills, and create confidence.

The Common Core math practice book provides numerous opportunities to evaluate basic skills along with abundant remediation and intervention activities. It is a skill that permits you to quickly master intricate information and produce better leads in less time.

Students can boost their test-taking skills by taking the book's two practice Common Core Math exams. All test questions answered and explained in detail.

Important Features of the 4th grade Common Core Math Book:

- A **complete review** of Common Core math test topics,
- Over 2,500 practice problems covering all topics tested,
- The most important concepts you need to know,
- Clear and concise, easy-to-follow sections,
- Well designed for enhanced learning and interest,
- Hands-on experience with all question types,
- **2 full-length practice tests** with detailed answer explanations,
- Cost-Effective Pricing,

Powerful math exercises to help you avoid traps and pacing yourself to beat the Common Core test. Students will gain valuable experience and raise their confidence by taking 4th grade math practice tests, learning about test structure, and gaining a deeper understanding of what is tested on the Common Core math grade 4. If ever there was a book to respond to the pressure to increase students' test scores, this is it.

WWW.MathNotion.COM

… So Much More Online!

- ✓ FREE Math Lessons
- ✓ More Math Learning Books!
- ✓ Mathematics Worksheets
- ✓ Online Math Tutors

For a PDF Version of This Book

SCAN ME

Please Visit WWW.MathNotion.com

Contents

Chapter 1 : Place Values and Number Sense

Topics that you'll learn in this chapter:

- ✓ Place Values,
- ✓ Compare and Ordering Numbers,
- ✓ Number in Standard Form
- ✓ Numbers in Word Form,
- ✓ Roman Numerals,
- ✓ Rounding Numbers,
- ✓ Odd or Even,

Place Values

Write numbers in expanded form.

1) Sixty–two ___ + ___

2) fifty–six ___ + ___

3) thirty–one ___ + ___

4) forty–five ___ + ___

5) twenty–eight ___ + ___

Circle the correct choice.

6) The 6 in 56 is in the

Ones place tens place hundreds place

7) The 2 in 25 is in the

Ones place tens place hundreds place

8) The 9 in 918 is in the

Ones place tens place hundreds place

9) The 3 in 537 is in the

Ones place tens place hundreds place

10) The 9 in 289 is in the

Ones place tens place hundreds place

Comparing and Ordering Numbers

Use less than, equal to or greater than.

1) 31 _______ 33
2) 57 _______ 49
3) 92 _______ 88
4) 76 _______ 67
5) 43 _______ 43
6) 54 _______ 46
7) 97 _______ 88
8) 42 _______ 36
9) 55 _______ 55
10) 57 _______ 75
11) 28 _______ 38
12) 19 _______ 15
13) 82 _______ 90
14) 78 _______ 84

Order each set numbers from least to greatest.

15) – 18, – 22, 28, – 17, 4 ___, ___, ___, ___, ___, ___

16) 19, –36, 11, – 12, 5 ___, ___, ___, ___, ___, ___

17) 27, – 56, 20, 1, – 27 ___, ___, ___, ___, ___, ___

18) 26, – 96, 2, – 26, 87, –75 ___, ___, ___, ___, ___, ___

19) –10, –71, 70, –26, –59, –39 ___, ___, ___, ___, ___, ___

20) 88, 4, 38, 7, 78, 9 ___, ___, ___, ___, ___, ___

21) 84, 14, 24, 0, 35, 22 ___, ___, ___, ___, ___, __

Numbers in Standard Form

Write the number in standard form.

1) 14 million 154 thousand 8

____

2) 89 million 15 thousand 798

____

3) 97 million 5 thousand 8

____

4) 124 million 2 thousand 2

____

5) 50 billion 3 million 5 thousand 4

____

6) 34 billion 45 million 578

____

7) 94 billion 21 million 51 thousand

____

8) 58 billion 708 thousand 120

____

9) 59 billion 54 thousand 86

____

10) 74 billion 354 thousand 158

____

11) 7 billion 13 million 12 thousand 7

____

12) 72 billion 450 million 658

____

13) 398 million 67 thousand 128

____

14) 24 billion 54 million 9 thousand 32

____

15) 795 million 458

____

16) 38 billion 2 million 54 thousand 9

____

Numbers in Word Form

Write each number in words.

1) 372 ____________________

2) 605 ____________________

3) 550 ____________________

4) 351 ____________________

5) 793 ____________________

6) 647 ____________________

7) 3,219 ____________________

8) 5,326 ____________________

9) 2,842 ____________________

10) 4,691____________________

11) 5,531____________________

12) 7,360____________________

13) 2,532____________________

14) 8,014____________________

15) 11,242____________________

Roman Numerals

Write in Romans numerals.

1	I	11	XI	21	XXI
2	II	12	XII	22	XXII
3	III	13	XIII	23	XXIII
4	IV	14	XIV	24	XXIV
5	V	15	XV	25	XXV
6	VI	16	XVI	26	XXVI
7	VII	17	XVII	27	XXVII
8	VIII	18	XVIII	28	XXVIII
9	IX	19	XIX	29	XXIX
10	X	20	XX	30	XXX

1) 11 __________ 2)21 __________

3) 24 __________ 4) 16 __________

5) 27 __________ 6) 29 __________

7) 12 __________ 8) 28 __________

9) 15 __________ 10) 20 __________

11) Add 16 + 14 and write in Roman numerals. __________

12) Subtract 34 – 5 and write in Roman numerals. __________

Rounding Numbers

Round each number to the underlined place value.

1) 3,<u>7</u>93

2) 3,<u>8</u>76

3) 3,4<u>5</u>2

4) 7,1<u>9</u>3

5) 5,2<u>7</u>8

6) 1,4<u>7</u>7

7) 8,<u>3</u>13

8) 24.<u>6</u>8

9) 8<u>4</u>.92

10) 71.<u>3</u>4

11) 66<u>4</u>.7

12) <u>9</u>,135

13) 15.3<u>8</u>1

14) 4,<u>5</u>21

15) 3<u>6</u>.50

16) 4,<u>8</u>19

17) 6,6<u>8</u>5

18) 2,5<u>3</u>8

19) 73.<u>6</u>2

20) 16,<u>5</u>27

21) 2<u>9</u>.720

22) 12,3<u>6</u>6

23) 31,<u>7</u>29

24) 7,8<u>3</u>8

Odd or Even

Identify whether each number is even or odd.

1) 18 ______

2) 27 ______

3) 21 ______

4) 17 ______

5) 67 ______

6) 76 ______

7) 80 ______

8) 53 ______

9) 58 ______

10) 98 ______

11) 49 ______

12) 113 ______

Circle the even number in each group.

13) 52, 11, 35, 73, 5, 29

14) 13, 15, 113, 87, 71, 18

15) 33, 45, 86, 59, 63, 87

16) 55, 32, 79, 51, 21, 83

Circle the odd number in each group.

17) 54, 36, 48, 76, 71, 100

18) 32, 56, 40, 74, 98, 67

19)58, 92, 25, 78, 76, 50

20)89, 12, 88, 42, 48, 120

Answers of Worksheets

Place Values

1) 60 + 2
2) 50 + 6
3) 30 + 1
4) 40 + 5
5) 20 + 8
6) ones place
7) tens place
8) hundreds place
9) tens place
10) ones place

Comparing and Ordering Numbers

1) 31 less than 33
2) 57 greater than 49
3) 92 greater than 88
4) 76 greater than 67
5) 43 equals to 43
6) 54 greater than 46
7) 97 greater than 88
8) 42 greater than 36
9) 55 equals to 55
10) 57 less than 75
11) 28 less than 38
12) 19 greater than 15
13) 82 less than 90
14) 78 less than 84
15) –22, –18, –17, 4, 28
16) –36, –12, 5, 11, 19
17) –56, –27, 1, 20, 27
18) –96, –75, –26, 2, 26, 87
19) –71, –59, –39, –26, –10, 70
20) 4, 7, 9, 38, 78, 88
21) 0, 14, 22,24, 35, 84

Numbers in Standard Form

1) 14,154,008
2) 89,015,798
3) 97,005,008
4) 124,002,002
5) 50,003,005,004
6) 34,054,000,578
7) 94,021,051,000
8) 58,000,708,120
9) 59,000,054,086
10) 74,000,354,158
11) 7,013,012,007
12) 72,450,000,658
13) 398,067,128
14) 24,054,009,032
15) 795,000,458
16) 38,002,054,009

Numbers in Word Form

1) three hundred seventy-two
2) six hundred five
3) five hundred fifty
4) three hundred fifty-one
5) seven hundred ninety -three
6) six hundred forty- seven
7) three thousand, two hundred nineteen
8) five thousand, three hundred twenty-six
9) two thousand, eight hundred forty-two
10) four thousand, six hundred ninety-one
11) five thousand, five hundred thirty-one
12) seven thousand, three hundred sixty
13) two thousand, five hundred thirty-two
14) eight thousand, fourteen

15) eleven thousand, two hundred forty-two

Roman Numerals

1) XI	4) XVI	7) XII	10) XX
2) XXI	5) XXVII	8) XXVIII	11) XXX
3) XXIV	6) XXIX	9) XV	12) XXIX

Rounding Numbers

1) 4,000	7) 8,300	13) 15.380	19) 73.60
2) 4,000	8) 24.70	14) 4,500	20) 16,500
3) 3,450	9) 85.00	15) 37.00	21) 30.00
4) 7,190	10) 71.30	16) 4,800	22) 12,370
5) 5,280	11) 665.00	17) 6,700	23) 31,700
6) 1,480	12) 9,000	18) 2,540	24) 7,840

Odd or Even

1) even	6) even	11) odd	16) 32
2) odd	7) even	12) odd	17) 71
3) odd	8) odd	13) 52	18) 67
4) odd	9) even	14) 18	19) 25
5) odd	10) even	15) 86	20) 89

Chapter 2 : Whole Number Operations

Topics that you'll learn in this chapter:

- ✓ Adding Whole Numbers,
- ✓ Subtracting Whole Numbers,
- ✓ Multiplying Whole Numbers,
- ✓ Dividing Hundreds,
- ✓ Long Division by One Digit,
- ✓ Division with Remainders,
- ✓ Rounding Whole Numbers,
- ✓ Whole Number Estimation,

Adding Whole Numbers

Add.

1) 5,763 + 8,238 = ______

2) 6,834 + 4,998 = ______

3) 3,548 + 5,693 = ______

4) 2,769 + 8,872 = ______

5) 3,196 + 2,936 = ______

6) 7,009 + 4,992 = ______

Find the missing numbers.

7) 3,468 + ___ = 4,102

8) 840 + 2,360 = ___

9) 5,200 + ___ = 7,980

10) 631 + ___ = 2,007

11) ___ + 803 = 3,945

12) ___ + 2,156 =5,922

13) David sells gems. He finds a diamond in Istanbul and buys it for $4,795. Then, he flies to Cairo and purchases a bigger diamond for the bargain price of $9,633. How much does David spend on the two diamonds? ____________

Subtracting Whole Numbers

Subtract.

1) $\begin{array}{r} 10{,}512 \\ -4{,}411 \\ \hline \end{array}$ ______

4) $\begin{array}{r} 8{,}001 \\ -5{,}224 \\ \hline \end{array}$ ______

2) $\begin{array}{r} 5{,}204 \\ -3{,}679 \\ \hline \end{array}$ ______

5) $\begin{array}{r} 11{,}916 \\ -8{,}711 \\ \hline \end{array}$ ______

3) $\begin{array}{r} 8{,}520 \\ -6{,}483 \\ \hline \end{array}$ ______

6) $\begin{array}{r} 5{,}005 \\ -2{,}008 \\ \hline \end{array}$ ______

Find the missing number.

7) 5,263 – ___ = 2,367

10) 6,511 – ___ = 3,759

8) 7,198 – ___ = 4,742

11)7,003 – 5,489 = ___

9) 8,928 –3,764 = ___

12) 8,800– 5,995 = ___

13) Jackson had $7,189 invested in the stock market until he lost $3,793 on those investments. How much money does he have in the stock market now?

____

Multiplying Whole Numbers

Find the answers.

1) $\begin{array}{r} 2{,}200 \\ \times\ 31 \\ \hline \end{array}$

2) $\begin{array}{r} 3{,}200 \\ \times\ 22 \\ \hline \end{array}$

3) $\begin{array}{r} 5{,}790 \\ \times\ 5 \\ \hline \end{array}$

4) $\begin{array}{r} 5{,}220 \\ \times\ 3 \\ \hline \end{array}$

5) $\begin{array}{r} 6{,}911 \\ \times\ 3 \\ \hline \end{array}$

6) $\begin{array}{r} 1{,}998 \\ \times 40 \\ \hline \end{array}$

7) $\begin{array}{r} 2{,}893 \\ \times 5.5 \\ \hline \end{array}$

8) $\begin{array}{r} 2{,}254 \\ \times\ 3.5 \\ \hline \end{array}$

9) $\begin{array}{r} 4{,}372 \\ \times\ 4.8 \\ \hline \end{array}$

10) $\begin{array}{r} 3{,}984 \\ \times\ 2.75 \\ \hline \end{array}$

11) $\begin{array}{r} 4{,}900 \\ \times\ 2.5 \\ \hline \end{array}$

12) $\begin{array}{r} 8{,}200 \\ \times\ 4.5 \\ \hline \end{array}$

Dividing Hundreds

Find answers.

1) 4,440 ÷ 400

2) 1,600 ÷ 40

3) 9,990 ÷ 90

4) 4,200 ÷ 60

5) 6,400 ÷ 8,000

6) 2,700 ÷ 30

7) 3,333 ÷ 30

8) 558 ÷ 45

9) 2,278 ÷ 85

10) 1,683 ÷ 55

11) 1,582 ÷ 35

12) 9,000 ÷ 600

13) 1,000 ÷ 2,500

14) 44.8 ÷ 20

15) 6,800 ÷ 400

16) 1,500 ÷ 5,000

17) 36.60 ÷ 120

18) 7,700 ÷ 700

19) 5,400 ÷ 600

20) 8,000 ÷ 160

21) 18,000 ÷ 9,000

22) 42,000 ÷ 30

23) 480 ÷ 40

24) 63,000 ÷ 900

Long Division by Two Digits

Find the quotient.

1) $18\overline{)576}$

2) $14\overline{)952}$

3) $21\overline{)588}$

4) $23\overline{)299}$

5) $44\overline{)748}$

6) $26\overline{)234}$

7) $16\overline{)496}$

8) $29\overline{)1,479}$

9) $54\overline{)1,080}$

10) $41\overline{)1,476}$

11) $53\overline{)2,491}$

12) $60\overline{)2,880}$

13) $32\overline{)2,912}$

14) $77\overline{)8,393}$

15) $85\overline{)3,740}$

16) $57\overline{)4,617}$

17) $50\overline{)9,200}$

18) $25\overline{)15,400}$

Division with Remainders

Find the quotient with remainder.

1) $14\overline{)715}$

2) $16\overline{)2,750}$

3) $27\overline{)4,603}$

4) $58\overline{)2,554}$

5) $42\overline{)7,732}$

6) $63\overline{)6,737}$

7) $71\overline{)9,036}$

8) $65\overline{)8,624}$

9) $35\overline{)5,705}$

10) $92\overline{)13,161}$

11) $46\overline{)12,214}$

12) $69\overline{)42,482}$

13) $85\overline{)6,858}$

14) $87\overline{)34,304}$

Rounding Whole Numbers

Round each number to the underlined place value.

1) 7,<u>5</u>33

2) 9,<u>3</u>74

3) 8,8<u>8</u>3

4) 2,3<u>6</u>8

5) 5,5<u>7</u>7

6) 3,3<u>8</u>1

7) 3,<u>5</u>20

8) 9,3<u>3</u>8

9) 8.<u>5</u>81

10) 33.<u>5</u>7

11) 51.<u>6</u>9

12) 22.<u>1</u>38

13) <u>6</u>,758

14) 11,5<u>5</u>7

15) 8,8<u>3</u>8

16) 5.<u>8</u>89

17) 1.<u>8</u>60

18) 25.<u>0</u>70

19) <u>9</u>.332

20) 49.<u>4</u>8

21) 28.<u>8</u>9

22) 24,3<u>7</u>7

23) 52,1<u>5</u>8

24) 13,8<u>8</u>3

25) 9,<u>6</u>09

26) 17,4<u>5</u>1

27) 18,<u>7</u>68

Whole Number Estimation

Estimate the sum by rounding each added to the nearest ten.

1) 875 + 325

2) 985 + 1,452

3) 2,424 + 4,128

4) 1,576 + 6,279

5) 1,247 + 3,863

6) 6,746 + 5,121

7) 3,924 + 6,456

8) 1,785 + 7,164

9) 1,458 + 2,442 = ______

10) 5,689 + 4,151 = ______

11) 8,259 + 4,754 = ______

12) 6,788 + 3,954 = ______

13) 9,123 + 4,455 = ______

14) 6,680 + 5,358 = ______

15) 3,165 + 7,124 = ______

16) 8,859 + 6,452 = ______

Answers of Worksheets

Adding Whole Numbers

1) 14,001
2) 11,832
3) 9,241
4) 11,641
5) 6,132
6) 12,001
7) 634
8) 3,200
9) 2,780
10) 1,376
11) 3,142
12) 3,766
13) $14,428

Subtracting Whole Numbers

1) 6,101
2) 1,525
3) 2,037
4) 2,777
5) 3,205
6) 2,997
7) 2,896
8) 2,456
9) 5,164
10) 2,752
11) 1,514
12) 2,805
13) 3,396

Multiplying Whole Numbers

1) 68,200
2) 70,400
3) 28,950
4) 15,660
5) 20,733
6) 79,920
7) 15,911.5
8) 7,889
9) 20,985.6
10) 10,956
11) 12,250
12) 36,900

Dividing Hundreds

1) 11.1
2) 40
3) 111
4) 70
5) 0.8
6) 90
7) 111.1
8) 12.4
9) 26.8
10) 30.6
11) 45.2
12) 15
13) 0.4
14) 2.24
15) 17
16) 0.3
17) 0.305
18) 11
19) 9
20) 50
21) 2
22) 1,400
23) 12
24) 70

Long Division by Two Digits

1) 32
2) 68
3) 28
4) 13
5) 17
6) 9
7) 31
8) 51
9) 20
10) 36
11) 47
12) 48
13) 91
14) 109
15) 44
16) 81
17) 184
18) 616

Division with Remainders

1) 51 R1
2) 171 R14
3) 170 R13
4) 44 R2
5) 184 R4
6) 106 R59
7) 127 R19
8) 132 R44
9) 163 R0
10) 143 R5
11) 265 R24
12) 615 R47
13) 80 R58
14) 394 R26

Rounding Whole Numbers

1) 7,500
2) 9,400
3) 8,880
4) 2,370
5) 5,580
6) 3,380
7) 3,500
8) 9,340
9) 8.60
10) 33.60
11) 51.70
12) 22.100
13) 7,000
14) 11,560
15) 8,840
16) 5.900
17) 1.900
18) 25.100
19) 9.000
20) 49.50
21) 28.90
22) 24,380
23) 52,160
24) 13,880
25) 9,600
26) 17,450
27) 18,800

Whole Number Estimation

1) 1,200
2) 2,440
3) 6,550
4) 7,860
5) 5,110
6) 11,870
7) 10,380
8) 8,950
9) 3,900
10) 9,840
11) 13,010
12) 10,740
13) 13,580
14) 12,040
15) 10,290
16) 15,310

Chapter 3 : Fractions

Topics that you'll learn in this chapter:

- ✓ Simplifying Fractions,
- ✓ Like Denominators,
- ✓ Compare Fractions with Like Denominators,
- ✓ More than two Fractions with Like Denominators,
- ✓ Unlike Denominators,
- ✓ Ordering Fractions,
- ✓ Denominators of 10, 100, and 1000,
- ✓ Multiply Fraction

Simplifying Fractions

Simplify the fractions.

1) $\frac{44}{84}$

2) $\frac{8}{20}$

3) $\frac{12}{16}$

4) $\frac{4}{24}$

5) $\frac{15}{30}$

6) $\frac{9}{63}$

7) $\frac{4}{14}$

8) $\frac{17}{51}$

9) $\frac{24}{30}$

10) $\frac{5}{35}$

11) $\frac{16}{48}$

12) $\frac{33}{22}$

13) $\frac{45}{63}$

14) $\frac{2.4}{3.2}$

15) $\frac{12}{60}$

16) $\frac{70}{112}$

17) $\frac{2.7}{7.2}$

18) $\frac{33}{88}$

19) $\frac{1.5}{13.5}$

20) $\frac{39}{52}$

21) $\frac{5}{45}$

22) $\frac{2.1}{4.2}$

Like Denominators

Add fractions.

1) $\frac{3}{4}+\frac{1}{4}$

2) $\frac{1}{5}+\frac{4}{5}$

3) $\frac{4}{9}+\frac{7}{9}$

4) $\frac{2}{7}+\frac{2}{7}$

5) $\frac{5}{13}+\frac{2}{13}$

6) $\frac{1}{14}+\frac{4}{14}$

7) $\frac{11}{19}+\frac{1}{19}$

8) $\frac{3}{16}+\frac{9}{16}$

9) $\frac{3}{10}+\frac{1}{10}$

10) $\frac{6}{17}+\frac{2}{17}$

11) $\frac{5}{22}+\frac{5}{22}$

12) $\frac{7}{35}+\frac{11}{35}$

13) $\frac{7}{27}+\frac{20}{27}$

14) $\frac{2}{31}+\frac{10}{31}$

15) $\frac{5}{23}+\frac{3}{23}$

16) $\frac{8}{41}+\frac{13}{41}$

17) $\frac{15}{37}+\frac{18}{37}$

18) $\frac{2}{51}+\frac{7}{51}$

19) $\frac{17}{26}+\frac{6}{26}$

20) $\frac{12}{48}+\frac{11}{48}$

21) $\frac{11}{29}+\frac{8}{29}$

22) $\frac{15}{34}+\frac{19}{34}$

23) $\frac{1}{19}+\frac{5}{19}$

24) $\frac{3}{53}+\frac{4}{53}$

25) $\frac{3}{20}+\frac{6}{20}$

26) $\frac{2}{63}+\frac{6}{63}$

27) $\frac{6}{38}+\frac{1}{38}$

28) $\frac{14}{31}+\frac{17}{31}$

29) $\frac{3}{28}+\frac{5}{28}$

30) $\frac{2}{37}+\frac{15}{37}$

Subtract fractions.

1) $\frac{8}{9} - \frac{4}{9}$

2) $\frac{3}{8} - \frac{1}{8}$

3) $\frac{9}{11} - \frac{3}{11}$

4) $\frac{9}{14} - \frac{4}{14}$

5) $\frac{15}{20} - \frac{8}{20}$

6) $\frac{8}{15} - \frac{7}{15}$

7) $\frac{11}{19} - \frac{9}{19}$

8) $\frac{13}{16} - \frac{1}{16}$

9) $\frac{7}{29} - \frac{4}{29}$

10) $\frac{14}{23} - \frac{7}{23}$

11) $\frac{15}{34} - \frac{7}{34}$

12) $\frac{18}{41} - \frac{9}{41}$

13) $\frac{17}{39} - \frac{16}{39}$

14) $\frac{6}{26} - \frac{2}{26}$

15) $\frac{14}{17} - \frac{4}{17}$

16) $\frac{33}{55} - \frac{20}{55}$

17) $\frac{41}{49} - \frac{36}{49}$

18) $\frac{40}{53} - \frac{39}{53}$

19) $\frac{27}{37} - \frac{17}{37}$

20) $\frac{21}{47} - \frac{11}{47}$

21) $\frac{24}{43} - \frac{12}{43}$

22) $\frac{13}{19} - \frac{12}{19}$

23) $\frac{6}{26} - \frac{3}{26}$

24) $\frac{9}{15} - \frac{7}{15}$

25) $\frac{8}{39} - \frac{3}{39}$

26) $\frac{18}{61} - \frac{15}{61}$

27) $\frac{12}{53} - \frac{9}{53}$

28) $\frac{75}{76} - \frac{74}{76}$

29) $\frac{26}{45} - \frac{13}{45}$

30) $\frac{20}{57} - \frac{17}{57}$

Compare Fractions with Like Denominators

Evaluate and compare. Write < or > or =.

1) $\frac{1}{3}+\frac{1}{3}$ ___ $\frac{1}{3}$

2) $\frac{3}{6}+\frac{3}{6}$ ___ $\frac{5}{6}$

3) $\frac{8}{9}-\frac{4}{9}$ ___ $\frac{7}{9}$

4) $\frac{4}{11}+\frac{5}{11}$ ___ $\frac{7}{11}$

5) $\frac{9}{14}-\frac{8}{14}$ ___ $\frac{5}{14}$

6) $\frac{11}{17}-\frac{3}{17}$ ___ $\frac{6}{17}$

7) $\frac{11}{21}+\frac{2}{21}$ ___ $\frac{10}{21}$

8) $\frac{8}{32}+\frac{6}{32}$ ___ $\frac{9}{32}$

9) $\frac{25}{29}-\frac{16}{29}$ ___ $\frac{11}{29}$

10) $\frac{28}{41}+\frac{13}{41}$ ___ $\frac{27}{41}$

11) $\frac{18}{35}-\frac{11}{35}$ ___ $\frac{22}{35}$

12) $\frac{32}{47}-\frac{22}{47}$ ___ $\frac{11}{47}$

13) $\frac{14}{27}+\frac{13}{27}$ ___ $\frac{24}{27}$

14) $\frac{34}{52}-\frac{11}{52}$ ___ $\frac{21}{52}$

15) $\frac{43}{56}-\frac{24}{56}$ ___ $\frac{27}{56}$

16) $\frac{27}{71}+\frac{25}{71}$ ___ $\frac{48}{71}$

More Than Two Fractions with Like Denominators

Add fractions.

1) $\frac{5}{9}+\frac{2}{9}+\frac{2}{9}$

2) $\frac{4}{6}+\frac{1}{6}+\frac{1}{6}$

3) $\frac{2}{17}+\frac{4}{17}+\frac{2}{17}$

4) $\frac{1}{5}+\frac{1}{5}+\frac{1}{5}$

5) $\frac{7}{18}+\frac{2}{18}+\frac{3}{18}$

6) $\frac{3}{27}+\frac{5}{27}+\frac{2}{27}$

7) $\frac{4}{33}+\frac{4}{33}+\frac{4}{33}$

8) $\frac{8}{23}+\frac{6}{23}+\frac{2}{23}$

9) $\frac{13}{41}+\frac{2}{41}+\frac{8}{41}$

10) $\frac{6}{35}+\frac{9}{35}+\frac{20}{35}$

11) $\frac{1}{37}+\frac{5}{37}+\frac{5}{37}$

12) $\frac{4}{43}+\frac{9}{43}+\frac{8}{43}$

13) $\frac{4}{51}+\frac{10}{51}+\frac{7}{51}$

14) $\frac{5}{26}+\frac{13}{26}+\frac{6}{26}$

15) $\frac{5}{64}+\frac{4}{64}+\frac{2}{64}$

16) $\frac{1}{73}+\frac{5}{73}+\frac{6}{73}$

Unlike Denominators

Add fraction.

1) $\frac{2}{9}+\frac{3}{4}$

2) $\frac{1}{4}+\frac{3}{5}$

3) $\frac{1}{16}+\frac{3}{4}$

4) $\frac{3}{8}+\frac{1}{7}$

5) $\frac{1}{3}+\frac{2}{4}$

6) $\frac{1}{6}+\frac{3}{7}$

7) $\frac{5}{18}+\frac{4}{6}$

8) $\frac{1}{12}+\frac{5}{6}$

9) $\frac{5}{27}+\frac{1}{9}$

10) $\frac{1}{6}+\frac{7}{24}$

11) $\frac{3}{5}+\frac{1}{8}$

12) $\frac{11}{42}+\frac{3}{7}$

13) $\frac{7}{20}+\frac{1}{3}$

14) $\frac{1}{45}+\frac{3}{5}$

15) $\frac{3}{32}+\frac{5}{8}$

16) $\frac{3}{48}+\frac{5}{6}$

17) $\frac{5}{12}+\frac{1}{6}$

18) $\frac{1}{34}+\frac{3}{17}$

19) $\frac{4}{9}+\frac{7}{54}$

20) $\frac{13}{56}+\frac{4}{7}$

21) $\frac{3}{12}+\frac{2}{3}$

22) $\frac{4}{33}+\frac{5}{11}$

Subtract fractions.

1) $\frac{8}{9}-\frac{1}{2}$

2) $\frac{2}{3}-\frac{3}{10}$

3) $\frac{1}{6}-\frac{1}{9}$

4) $\frac{7}{8}-\frac{1}{4}$

5) $\frac{3}{4}-\frac{1}{28}$

6) $\frac{11}{30}-\frac{3}{15}$

7) $\frac{11}{18}-\frac{5}{9}$

8) $\frac{5}{13}-\frac{3}{26}$

9) $\frac{17}{35}-\frac{2}{7}$

10) $\frac{5}{6}-\frac{12}{36}$

11) $\frac{5}{9}-\frac{1}{27}$

12) $\frac{3}{5}-\frac{1}{8}$

13) $\frac{2}{3}-\frac{3}{5}$

14) $\frac{7}{8}-\frac{3}{7}$

15) $\frac{5}{9}-\frac{13}{45}$

16) $\frac{3}{4}-\frac{5}{36}$

17) $\frac{39}{49}-\frac{5}{7}$

18) $\frac{3}{11}-\frac{3}{22}$

19) $\frac{17}{48}-\frac{4}{12}$

20) $\frac{2}{3}-\frac{4}{13}$

21) $\frac{5}{8}-\frac{19}{72}$

22) $\frac{3}{5}-\frac{1}{12}$

Ordering Fractions

Order the fractions from least to greatest.

1) $\frac{1}{5}, \frac{1}{11}, \frac{1}{8}, \frac{1}{3}$ ____, ____, ____, ____

2) $\frac{1}{9}, \frac{1}{18}, \frac{2}{4}, \frac{1}{5}$ ____, ____, ____, ____

3) $\frac{4}{7}, \frac{1}{7}, \frac{6}{21}, \frac{15}{21}$ ____, ____, ____, ____

4) $\frac{1}{2}, \frac{1}{3}, \frac{4}{9}, \frac{5}{18}$ ____, ____, ____, ____

5) $\frac{4}{9}, \frac{3}{4}, \frac{7}{36}, \frac{1}{6}$ ____, ____, ____, ____

Order the fractions from greatest to least.

6) $\frac{3}{4}, \frac{4}{7}, \frac{3}{10}, \frac{5}{13}$ ____, ____, ____, ____

7) $\frac{5}{11}, \frac{5}{6}, \frac{2}{5}, \frac{1}{3}$ ____, ____, ____, ____

8) $\frac{7}{8}, \frac{1}{6}, \frac{3}{4}, \frac{5}{15}$ ____, ____, ____, ____

9) $\frac{4}{7}, \frac{2}{3}, \frac{11}{25}, \frac{13}{33}$ ____, ____, ____, ____

10) $\frac{18}{20}, \frac{15}{16}, \frac{14}{18}, \frac{5}{12}$ ____, ____, ____, ____

Denominators of 10, 100, and 1000

Add fractions.

1) $\frac{7}{10} + \frac{13}{100}$

2) $\frac{1}{10} + \frac{10}{100}$

3) $\frac{15}{100} + \frac{1}{1,000}$

4) $\frac{56}{100} + \frac{3}{10}$

5) $\frac{50}{1,000} + \frac{7}{10}$

6) $\frac{6}{10} + \frac{30}{1,000}$

7) $\frac{9}{100} + \frac{3}{10}$

8) $\frac{5}{10} + \frac{50}{100}$

9) $\frac{48}{100} + \frac{6}{10}$

10) $\frac{70}{100} + \frac{2}{10}$

11) $\frac{80}{100} + \frac{200}{1,000}$

12) $\frac{30}{100} + \frac{4}{10}$

13) $\frac{9}{100} + \frac{7}{10}$

14) $\frac{25}{100} + \frac{6}{10}$

15) $\frac{15}{100} + \frac{8}{10}$

16) $\frac{3}{10} + \frac{31}{100}$

17) $\frac{8}{10} + \frac{11}{100}$

18) $\frac{34}{100} + \frac{6}{10}$

Subtract fractions.

1) $\frac{8}{10} - \frac{20}{100}$

2) $\frac{5}{10} - \frac{47}{100}$

3) $\frac{12}{100} - \frac{60}{1,000}$

4) $\frac{6}{10} - \frac{50}{100}$

5) $\frac{3}{10} - \frac{23}{100}$

6) $\frac{70}{100} - \frac{250}{1,000}$

7) $\frac{4}{10} - \frac{350}{1,000}$

8) $\frac{70}{100} - \frac{3}{10}$

9) $\frac{40}{100} - \frac{3}{10}$

10) $\frac{6}{10} - \frac{180}{1,000}$

11) $\frac{93}{100} - \frac{5}{10}$

12) $\frac{65}{100} - \frac{4}{10}$

13) $\frac{80}{100} - \frac{6}{10}$

14) $\frac{90}{100} - \frac{5}{10}$

15) $\frac{200}{1,000} - \frac{1}{10}$

16) $\frac{90}{100} - \frac{7}{10}$

17) $\frac{900}{1,000} - \frac{40}{100}$

18) $\frac{60}{100} - \frac{3}{10}$

Multiplying Fractions

Find the product.

1) $\frac{4}{5} \times \frac{2}{6} =$

2) $\frac{4}{22} \times \frac{5}{8} =$

3) $\frac{8}{30} \times \frac{12}{16} =$

4) $\frac{9}{14} \times \frac{21}{36} =$

5) $\frac{14}{15} \times \frac{5}{7} =$

6) $\frac{16}{19} \times \frac{3}{4} =$

7) $\frac{4}{9} \times \frac{9}{8} =$

8) $\frac{47}{85} \times 0 =$

9) $\frac{5}{8} \times \frac{16}{6} =$

10) $\frac{28}{15} \times \frac{5}{7} =$

11) $\frac{32}{24} \times \frac{12}{16} =$

12) $\frac{6}{42} \times \frac{7}{36} =$

13) $\frac{13}{8} \times \frac{12}{4} =$

14) $\frac{10}{9} \times \frac{6}{5} =$

15) $\frac{35}{56} \times \frac{8}{7} =$

16) $\frac{16}{18} \times 9 =$

17) $\frac{5}{22} \times \frac{44}{15} =$

18) $\frac{10}{18} \times \frac{9}{20} =$

19) $\frac{7}{11} \times \frac{8}{21} =$

20) $\frac{26}{24} \times \frac{8}{52} =$

21) $\frac{6}{17} \times \frac{1}{12} =$

22) $\frac{20}{9} \times \frac{6}{100} =$

23) $\frac{8}{14} \times \frac{7}{72} =$

24) $\frac{50}{100} \times \frac{300}{400} =$

Answers of Worksheets

Simplifying Fractions

1) $\frac{11}{21}$
2) $\frac{2}{5}$
3) $\frac{3}{4}$
4) $\frac{1}{6}$
5) $\frac{1}{2}$
6) $\frac{1}{7}$
7) $\frac{2}{7}$
8) $\frac{1}{3}$
9) $\frac{4}{5}$
10) $\frac{1}{7}$
11) $\frac{1}{3}$
12) $\frac{3}{2}$
13) $\frac{5}{7}$
14) $\frac{3}{4}$
15) $\frac{1}{5}$
16) $\frac{5}{8}$
17) $\frac{3}{8}$
18) $\frac{3}{8}$
19) $\frac{1}{9}$
20) $\frac{3}{4}$
21) $\frac{1}{9}$
22) $\frac{1}{2}$

Like Denominators (addition)

1) 1
2) 1
3) $\frac{11}{9}$
4) $\frac{4}{7}$
5) $\frac{7}{13}$
6) $\frac{5}{14}$
7) $\frac{12}{19}$
8) $\frac{3}{4}$
9) $\frac{2}{5}$
10) $\frac{8}{17}$
11) $\frac{5}{11}$
12) $\frac{18}{35}$
13) 1
14) $\frac{12}{31}$
15) $\frac{8}{23}$
16) $\frac{21}{41}$
17) $\frac{33}{37}$
18) $\frac{3}{17}$
19) $\frac{23}{26}$
20) $\frac{23}{48}$
21) $\frac{19}{29}$
22) 1
23) $\frac{6}{19}$
24) $\frac{7}{53}$
25) $\frac{9}{20}$
26) $\frac{8}{63}$
27) $\frac{7}{38}$
28) 1
29) $\frac{2}{7}$
30) $\frac{17}{37}$

Like Denominators (Subtraction)

1) $\frac{4}{9}$
2) $\frac{1}{4}$
3) $\frac{6}{11}$
4) $\frac{5}{14}$
5) $\frac{7}{20}$
6) $\frac{1}{15}$
7) $\frac{2}{19}$
8) $\frac{3}{4}$
9) $\frac{3}{29}$
10) $\frac{7}{23}$
11) $\frac{8}{34}$
12) $\frac{9}{41}$
13) $\frac{1}{39}$
14) $\frac{2}{13}$
15) $\frac{10}{17}$
16) $\frac{13}{55}$
17) $\frac{5}{49}$
18) $\frac{1}{53}$
19) $\frac{10}{37}$
20) $\frac{10}{47}$
21) $\frac{12}{43}$
22) $\frac{1}{19}$
23) $\frac{3}{26}$
24) $\frac{2}{15}$
25) $\frac{5}{39}$
26) $\frac{3}{61}$
27) $\frac{3}{53}$
28) $\frac{1}{76}$

29) $\frac{13}{45}$

30) $\frac{1}{19}$

Compare Fractions with Like Denominators

1) $\frac{2}{3} > \frac{1}{3}$
2) $1 > \frac{5}{6}$
3) $\frac{4}{9} < \frac{7}{9}$
4) $\frac{9}{11} > \frac{7}{11}$
5) $\frac{1}{14} < \frac{5}{14}$
6) $\frac{8}{17} > \frac{6}{17}$
7) $\frac{13}{21} > \frac{10}{21}$
8) $\frac{14}{32} > \frac{9}{32}$
9) $\frac{9}{29} < \frac{11}{29}$
10) $1 > \frac{27}{41}$
11) $\frac{7}{35} < \frac{22}{35}$
12) $\frac{10}{47} < \frac{11}{47}$
13) $1 > \frac{24}{27}$
14) $\frac{23}{52} > \frac{21}{52}$
15) $\frac{19}{56} < \frac{27}{56}$
16) $\frac{52}{71} > \frac{48}{71}$

More Than Two Fractions with Like Denominators

1) 1
2) 1
3) $\frac{8}{17}$
4) $\frac{3}{5}$
5) $\frac{2}{3}$
6) $\frac{10}{27}$
7) $\frac{4}{11}$
8) $\frac{16}{23}$
9) $\frac{23}{41}$
10) 1
11) $\frac{11}{37}$
12) $\frac{21}{43}$
13) $\frac{7}{17}$
14) $\frac{12}{13}$
15) $\frac{11}{64}$
16) $\frac{12}{73}$

Unlike Denominators (Addition)

1) $\frac{35}{36}$
2) $\frac{17}{20}$
3) $\frac{13}{16}$
4) $\frac{29}{56}$
5) $\frac{5}{6}$
6) $\frac{25}{42}$
7) $\frac{17}{18}$
8) $\frac{11}{12}$
9) $\frac{8}{27}$
10) $\frac{11}{24}$
11) $\frac{29}{40}$
12) $\frac{29}{42}$
13) $\frac{41}{60}$
14) $\frac{28}{45}$
15) $\frac{23}{32}$
16) $\frac{43}{48}$
17) $\frac{7}{12}$
18) $\frac{7}{34}$
19) $\frac{31}{54}$
20) $\frac{45}{56}$
21) $\frac{11}{12}$
22) $\frac{19}{33}$

Unlike Denominators (Subtraction)

1) $\frac{7}{18}$
2) $\frac{11}{30}$
3) $\frac{1}{18}$
4) $\frac{5}{8}$
5) $\frac{5}{7}$
6) $\frac{1}{6}$
7) $\frac{1}{18}$
8) $\frac{7}{26}$
9) $\frac{1}{5}$
10) $\frac{1}{2}$
11) $\frac{14}{27}$
12) $\frac{19}{40}$
13) $\frac{1}{15}$
14) $\frac{25}{56}$
15) $\frac{4}{15}$
16) $\frac{11}{18}$
17) $\frac{4}{49}$
18) $\frac{3}{22}$
19) $\frac{1}{48}$
20) $\frac{14}{39}$
21) $\frac{13}{36}$
22) $\frac{31}{60}$

Ordering Fractions

1) $\frac{1}{11}, \frac{1}{8}, \frac{1}{5}, \frac{1}{3}$

2) $\frac{1}{18}, \frac{1}{9}, \frac{1}{5}, \frac{2}{4},$

3) $\frac{1}{7}, \frac{6}{21}, \frac{4}{7}, \frac{15}{21}$

4) $\frac{5}{18}, \frac{1}{3}, \frac{4}{9}, \frac{1}{2}$

5) $\frac{1}{6}, \frac{7}{36}, \frac{4}{9}, \frac{3}{4}$

6) $\frac{3}{4}, \frac{4}{7}, \frac{5}{13}, \frac{3}{10}$

7) $\frac{5}{6}, \frac{5}{11}, \frac{2}{5}, \frac{1}{3}$

8) $\frac{7}{8}, \frac{3}{4}, \frac{5}{15}, \frac{1}{6}$

9) $\frac{2}{3}, \frac{4}{7}, \frac{11}{25}, \frac{13}{33}$

10) $\frac{15}{16}, \frac{18}{20}, \frac{14}{18}, \frac{5}{12}$

Denominators of 10, 100, and 1000

1) $\frac{83}{100}$

2) $\frac{1}{5}$

3) $\frac{151}{1,000}$

4) $\frac{43}{50}$

5) $\frac{3}{4}$

6) $\frac{63}{100}$

7) $\frac{39}{100}$

8) 1

9) $\frac{27}{25}$

10) $\frac{9}{10}$

11) 1

12) $\frac{7}{10}$

13) $\frac{79}{100}$

14) $\frac{17}{20}$

15) $\frac{19}{20}$

16) $\frac{61}{100}$

17) $\frac{91}{100}$

18) $\frac{47}{50}$

Denominators of 10, 100, and 1000 (Subtract)

1) $\frac{3}{5}$

2) $\frac{3}{100}$

3) $\frac{3}{50}$

4) $\frac{1}{10}$

5) $\frac{7}{100}$

6) $\frac{9}{20}$

7) $\frac{1}{20}$

8) $\frac{2}{5}$

9) $\frac{1}{10}$

10) $\frac{21}{50}$

11) $\frac{43}{100}$

12) $\frac{1}{4}$

13) $\frac{1}{5}$

14) $\frac{2}{5}$

15) $\frac{1}{10}$

16) $\frac{1}{5}$

17) $\frac{1}{2}$

18) $\frac{3}{10}$

Multiplying Fractions

1) $\frac{4}{15}$

2) $\frac{5}{44}$

3) $\frac{1}{5}$

4) $\frac{3}{8}$

5) $\frac{2}{3}$

6) $\frac{12}{19}$

7) $\frac{1}{2}$

8) 0

9) $\frac{5}{3}$

10) $\frac{4}{3}$

11) 1

12) $\frac{1}{36}$

13) $\frac{39}{8}$

14) $\frac{4}{3}$

15) $\frac{5}{7}$

16) 8

17) $\frac{2}{3}$

18) $\frac{1}{4}$

19) $\frac{8}{33}$

20) $\frac{1}{6}$

21) $\frac{1}{34}$

22) $\frac{2}{15}$

23) $\frac{1}{18}$

24) $\frac{3}{8}$

Chapter 4 :

Mixed Numbers

Topics that you'll learn in this chapter:

- ✓ Fractions to Mixed Numbers,
- ✓ Mixed Numbers to Fractions,
- ✓ Add and Subtract Mixed Numbers,
- ✓ Multiply Mixed Numbers

Fractions to Mixed Numbers

Convert fractions to mixed numbers.

1) $\frac{9}{5}$

2) $\frac{11}{3}$

3) $\frac{39}{8}$

4) $\frac{27}{11}$

5) $\frac{7}{2}$

6) $\frac{43}{4}$

7) $\frac{49}{9}$

8) $\frac{15}{4}$

9) $\frac{37}{7}$

10) $\frac{19}{7}$

11) $\frac{41}{9}$

12) $\frac{45}{12}$

13) $\frac{17}{5}$

14) $\frac{29}{6}$

15) $\frac{13}{4}$

16) $\frac{15}{7}$

17) $\frac{65}{7}$

18) $\frac{59}{8}$

19) $\frac{25}{4}$

20) $\frac{17}{8}$

Mixed Numbers to Fractions

Convert to fraction.

1) $3\frac{3}{5}$

2) $1\frac{1}{3}$

3) $4\frac{2}{5}$

4) $4\frac{2}{8}$

5) $2\frac{1}{5}$

6) $2\frac{8}{11}$

7) $4\frac{4}{7}$

8) $3\frac{7}{12}$

9) $2\frac{1}{3}$

10) $7\frac{5}{7}$

11) $2\frac{7}{10}$

12) $3\frac{4}{9}$

13) $1\frac{5}{8}$

14) $4\frac{3}{11}$

15) $3\frac{4}{7}$

16) $5\frac{2}{8}$

17) $7\frac{1}{7}$

18) $13\frac{1}{2}$

19) $4\frac{2}{7}$

20) $5\frac{2}{10}$

21) $12\frac{1}{3}$

22) $7\frac{1}{8}$

Add and Subtract Mixed Numbers

Add mixed numbers.

1) $3\frac{2}{5} + 8\frac{1}{5}$

2) $3\frac{2}{3} + 4\frac{1}{2}$

3) $6\frac{2}{7} + 2\frac{3}{7}$

4) $4\frac{2}{5} + 3\frac{1}{4}$

5) $8\frac{3}{4} - 2\frac{1}{2}$

6) $6\frac{5}{12} - 4\frac{1}{4}$

7) $5\frac{3}{8} - 3\frac{7}{8}$

8) $6\frac{1}{4} - 2\frac{15}{16}$

9) $9\frac{23}{28} - 4\frac{17}{28}$

10) $6\frac{1}{6} + 6\frac{2}{3}$

11) $4\frac{2}{9} + 5\frac{5}{9}$

12) $2\frac{1}{4} + 7\frac{4}{7}$

13) $7\frac{1}{5} - 3\frac{3}{5}$

14) $3\frac{1}{6} + 2\frac{3}{7}$

15) $2\frac{1}{3} + 4\frac{1}{4}$

16) $4\frac{1}{4} - 1\frac{2}{5}$

17) $\frac{1}{3} + 6\frac{1}{6}$

18) $2\frac{3}{5} + 2\frac{1}{10}$

Multiplying Mixed Number

Multiply. Reduce to lowest terms.

1) $2\frac{3}{5} \times 1\frac{3}{4} =$

2) $1\frac{5}{6} \times 1\frac{1}{3} =$

3) $2\frac{3}{5} \times 1\frac{1}{7} =$

4) $3\frac{1}{7} \times 2\frac{1}{2} =$

5) $4\frac{3}{4} \times 1\frac{1}{4} =$

6) $3\frac{1}{2} \times 1\frac{4}{5} =$

7) $3\frac{3}{4} \times 1\frac{1}{2} =$

8) $5\frac{2}{3} \times 3\frac{1}{3} =$

9) $3\frac{2}{3} \times 3\frac{1}{2} =$

10) $2\frac{1}{3} \times 3\frac{1}{2} =$

11) $4\frac{3}{4} \times 3\frac{2}{3} =$

12) $3\frac{2}{4} \times 3\frac{1}{6} =$

13) $2\frac{2}{5} \times 1\frac{1}{3} =$

14) $2\frac{1}{3} \times 1\frac{1}{6} =$

15) $2\frac{2}{3} \times 3\frac{1}{2} =$

16) $2\frac{1}{8} \times 2\frac{2}{5} =$

17) $2\frac{1}{4} \times 1\frac{2}{3} =$

18) $2\frac{3}{5} \times 1\frac{1}{4} =$

19) $2\frac{3}{5} \times 1\frac{5}{6} =$

20) $3\frac{3}{5} \times 2\frac{3}{4} =$

21) $3\frac{3}{4} \times 1\frac{1}{3} =$

22) $2\frac{5}{8} \times 3\frac{1}{4} =$

Answers of Worksheets

Fractions to Mixed Numbers

1) $1\frac{4}{5}$
2) $3\frac{2}{3}$
3) $4\frac{7}{8}$
4) $2\frac{5}{11}$
5) $3\frac{1}{2}$
6) $10\frac{3}{4}$
7) $5\frac{4}{9}$
8) $3\frac{3}{4}$
9) $5\frac{2}{7}$
10) $2\frac{5}{7}$
11) $4\frac{5}{9}$
12) $3\frac{9}{12}$
13) $3\frac{2}{5}$
14) $4\frac{5}{6}$
15) $3\frac{1}{4}$
16) $2\frac{1}{7}$
17) $9\frac{2}{7}$
18) $7\frac{3}{8}$
19) $6\frac{1}{4}$
20) $2\frac{1}{8}$

Mixed Numbers to Fractions

1) $\frac{18}{5}$
2) $\frac{4}{3}$
3) $\frac{22}{5}$
4) $\frac{34}{8}$
5) $\frac{11}{5}$
6) $\frac{30}{11}$
7) $\frac{32}{7}$
8) $\frac{43}{12}$
9) $\frac{7}{3}$
10) $\frac{54}{7}$
11) $\frac{27}{10}$
12) $\frac{31}{9}$
13) $\frac{13}{8}$
14) $\frac{47}{11}$
15) $\frac{25}{7}$
16) $\frac{42}{8}$
17) $\frac{50}{7}$
18) $\frac{27}{2}$
19) $\frac{30}{7}$
20) $\frac{52}{10}$
21) $\frac{37}{3}$
22) $\frac{57}{8}$

Add and Subtract Mixed Numbers

1) $11\frac{3}{5}$
2) $8\frac{1}{6}$
3) $8\frac{5}{7}$
4) $7\frac{13}{20}$
5) $6\frac{1}{4}$
6) $2\frac{1}{6}$
7) $1\frac{1}{2}$
8) $3\frac{5}{16}$
9) $5\frac{3}{14}$
10) $12\frac{5}{6}$
11) $9\frac{7}{9}$
12) $9\frac{23}{28}$
13) $3\frac{3}{5}$
14) $5\frac{25}{42}$
15) $6\frac{7}{12}$
16) $2\frac{17}{20}$
17) $6\frac{1}{2}$
18) $4\frac{7}{10}$

Multiplying Mixed Number

1) $4\frac{11}{20}$
2) $2\frac{4}{9}$
3) $2\frac{34}{35}$
4) $7\frac{6}{7}$
5) $5\frac{15}{16}$
6) $6\frac{3}{10}$
7) $5\frac{5}{8}$
8) $18\frac{8}{9}$
9) $12\frac{5}{6}$

10) $8\frac{1}{6}$

11) $17\frac{5}{12}$

12) $11\frac{1}{12}$

13) $3\frac{1}{5}$

14) $2\frac{13}{18}$

15) $9\frac{1}{3}$

16) $5\frac{1}{10}$

17) $3\frac{3}{4}$

18) $3\frac{1}{4}$

19) $4\frac{23}{30}$

20) $9\frac{9}{10}$

21) 5

22) $8\frac{17}{32}$

Chapter 5 : Decimals

Topics that you'll learn in this chapter:

- ✓ Graph Decimal
- ✓ Adding and Subtracting Decimals,
- ✓ Multiplying and Dividing Decimals,
- ✓ Round decimals,
- ✓ Comparing Decimals,

Graph Decimals

Write the decimals indicated by the arrows.

1)

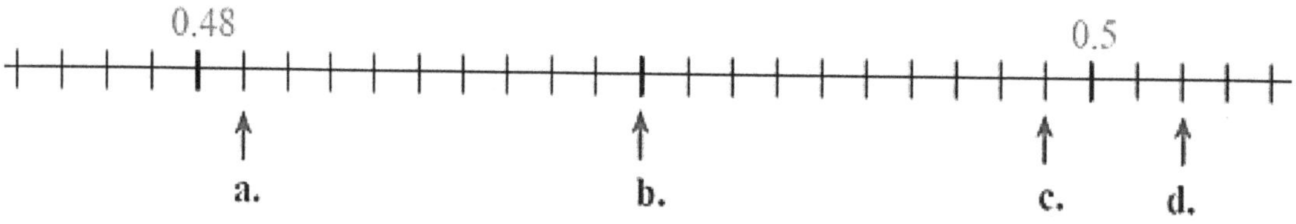

a. __________ b. __________ c. __________ d. __________

2)

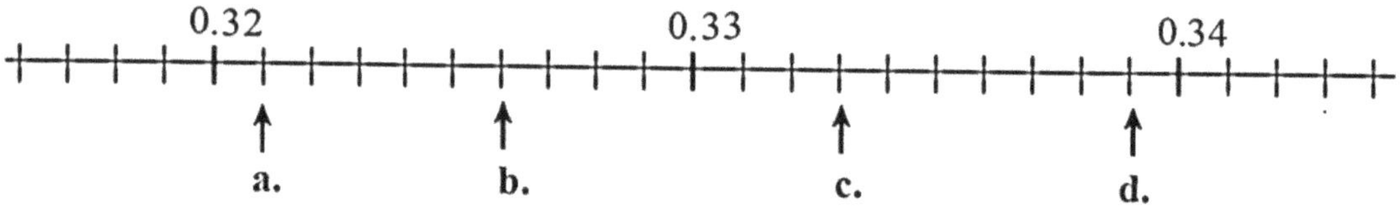

a. __________ b. __________ c. __________ d. __________

3)

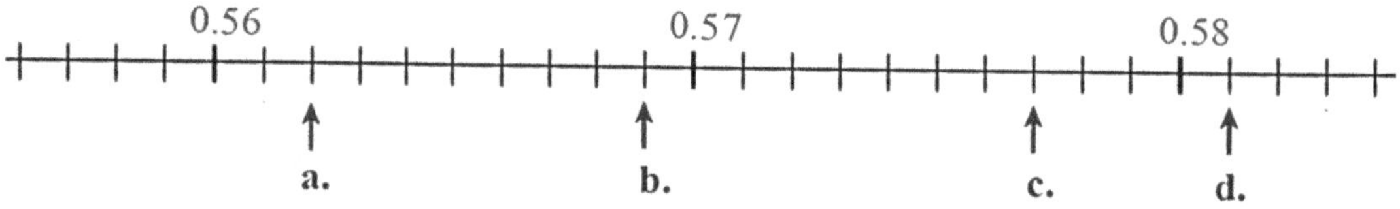

a. __________ b. __________ c. __________ d. __________

4)

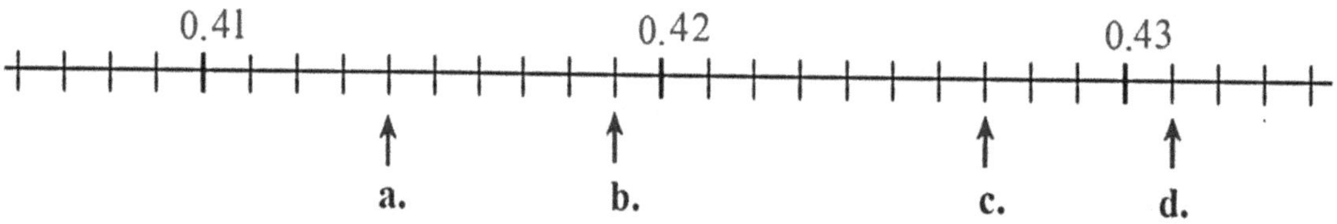

a. __________ b. __________ c. __________ d. __________

Adding and Subtracting Decimals

Add and subtract decimals.

1) $\begin{array}{r} 24.19 \\ -\ 15.42 \\ \hline \end{array}$ ____

2) $\begin{array}{r} 42.23 \\ +\ 25.42 \\ \hline \end{array}$ ____

3) $\begin{array}{r} 72.54 \\ +\ 11.28 \\ \hline \end{array}$ ____

4) $\begin{array}{r} 57.45 \\ -\ 24.75 \\ \hline \end{array}$ ____

5) $\begin{array}{r} 43.57 \\ +\ 54.85 \\ \hline \end{array}$ ____

6) $\begin{array}{r} 86.68 \\ -\ 54.12 \\ \hline \end{array}$ ____

Solve.

7) ____ + 2.7 = 8.1

8) 6.4 + ____ = 12.8

9) 7.9 + ____ = 17

10) 4.6 + ____ = 15.3

11) ____ + 9.4 = 15

12) ____ + 8.24 = 13.54

Order each set of numbers from least to greatest.

1) 0.4, 0.67, 0.44, 0.73, 0.51 ___, ___, ___, ___, ___, ___

2) 3.9, 6.1, 4.28, 7.02, 4.65 ___, ___, ___, ___, ___, ___

3) 1.9, 1.04, 0.79, 0.72, 0.09 ___, ___, ___, ___, ___, ___

4) 2.6, 5.2, 1.9, 4.01, 1.99, 3.2 ___, ___, ___, ___, ___, ___

5) 4.2, 6.1, 3.8, 5.7, 2.1, 2.8 ___, ___, ___, ___, ___, ___

6) 0.56, 0.87, 0.14, 1.24, 3.1 ___, ___, ___, ___, ___, ___

Multiplying and Dividing Decimals

Find each product.

1) $\begin{array}{r} 1.5 \\ \times\ 2.1 \\ \hline \end{array}$

4) $\begin{array}{r} 6.5 \\ \times\ 0.99 \\ \hline \end{array}$

7) $\begin{array}{r} 4.8 \\ \times\ 9.1 \\ \hline \end{array}$

2) $\begin{array}{r} 4.6 \\ \times\ 3.4 \\ \hline \end{array}$

5) $\begin{array}{r} 12.1 \\ \times\ 5.2 \\ \hline \end{array}$

8) $\begin{array}{r} 22.35 \\ \times\ 20 \\ \hline \end{array}$

3) $\begin{array}{r} 6.3 \\ \times\ 2.5 \\ \hline \end{array}$

6) $\begin{array}{r} 3.4 \\ \times\ 8.9 \\ \hline \end{array}$

9) $\begin{array}{r} 15.25 \\ \times\ 3.6 \\ \hline \end{array}$

Find each quotient.

10) $3.5 \div 0.85$

11) $15.35 \div 4.6$

12) $32.42 \div 8.8$

13) $9.2 \div 3.4$

14) $0.84 \div 0.1$

15) $21.5 \div 1{,}000$

16) $4.1 \div 100$

17) $9.7 \div 10$

18) $6.55 \div 1.25$

19) $18.48 \div 11.2$

Rounding Decimals

Round each decimal number to the nearest place indicated.

1) 0.$\underline{3}$2
2) 5.$\underline{0}$1
3) 8.$\underline{8}$24
4) 0.$\underline{4}$78
5) $\underline{7}$.32
6) 0.$\underline{2}$9
7) 11.$\underline{3}$1
8) $\underline{6}$.223
9) 9.6$\underline{3}$7
10) 5.$\underline{4}$804
11) $\underline{7}$.9
12) $\underline{5}$.2439
13) 6.$\underline{4}$92
14) 1.$\underline{6}$2
15) 7$\underline{2}$.85
16) 8$\underline{3}$.67
17) 41.$\underline{6}$8
18) 79$\underline{4}$.741
19) 5$\underline{2}$.2
20) 7$\underline{6}$.93
21) $\underline{3}$.219
22) 7$\underline{2}$.09
23) 486.$\underline{4}$91
24) 7.$\underline{0}$8

Comparing Decimals

Write the correct comparison symbol (>, < or =).

1) 0.35....1.8

2) 1.9......1.19

3) 8.6......8.6

4) 2.45....24.5

5) 7.56....0.756

6) 11.4....11.05

7) 7.4......0.7.4

8) 8.56....0.85

9) 7..........0.7

10) 7.12....0.712

11) 12.3....12.5

12) 4.67....4.68

13) 2.57....2.75

14) 3.46....0.346

15) 6.87....6.78

16) 0.89....0.98

17) 1.57....0.157

18) 0.092..0.091

19) 24.3....24.3

20) 0.17....0.71

21) 0.46....0.64

22) 0.2......0.08

23) 0.10....0.1

24) 3.52....31.5

Answers of Worksheets

Graph Decimals

1) a. 0.481 b. 0.49 c. 0.499 d. 0.502
2) a. 0.321 b. 0.326 c. 0.333 d. 0339
3) a. 0.562 b. 0.569 c. 0.577 d. 0.581
4) a. 0.414 b. 0.419 c. 0.427 d. 0431

Adding and Subtracting Decimals

1) 8.77
2) 67.65
3) 83.82
4) 32.7
5) 98.42
6) 32.56
7) 5.4
8) 6.4
9) 9.1
10) 10.7
11) 5.6
12) 5.3

Order and Comparing Decimals

1) 0.4, 0.44, 0.51, 0.67, 0.73
2) 3.9, 4.28, 4.65, 6.1, 7.02
3) 0.09, 0.72, 0.79, 1.04, 1.9
4) 1.9, 1.99, 2.6, 3.2, 4.01, 5.2
5) 2.1, 2.8, 3.8, 4.2, 5.7, 6.1
6) 0.14, 0.56, 0.87, 1.24, 3.1

Multiplying and Dividing Decimals

1) 3.15
2) 15.64
3) 15.75
4) 6.435
5) 62.92
6) 30.26
7) 43.68
8) 447
9) 54.9
10) 4.117…
11) 3.336…
12) 3.684…
13) 2.705…
14) 8.4
15) 0.0215
16) 0.041
17) 0.97
18) 5.24
19) 1.65

Rounding Decimals

1) 0.3
2) 5.0
3) 8.8
4) 0.5
5) 7
6) 0.3
7) 11.3
8) 6
9) 9.64
10) 5.5
11) 8
12) 5
13) 6.5
14) 1.6
15) 73
16) 84
17) 41.7
18) 795
19) 52
20) 77
21) 3
22) 72
23) 486.5
24) 7.1

Comparing Decimals

1) 0.35 < 1.8
2) 1.9 > 1.19
3) 8.6 = 8.6
4) 2.45 < 24.5
5) 7.56 > 0.756
6) 11.4 > 11.05
7) 7.4 > 0.74
8) 8.56 > 0.85
9) 7 > 0.7
10) 7.12 > 0.712
11) 12.3 < 12.5
12) 4.67 < 4.68

13) $2.57 < 2.75$

14) $3.46 > 0.346$

15) $6.87 > 6.78$

16) $0.89 < 0.98$

17) $1.57 > 0.157$

18) $0.092 > 0.091$

19) $24.3 = 24.3$

20) $0.17 < 0.71$

21) $0.46 < 0.64$

22) $0.2 > 0.08$

23) $0.10 = 0.1$

24) $3.52 < 31.5$

Chapter 6 : Patterns and Algebraic Thinking

Topics that you'll learn in this chapter:

- ✓ Repeating Patterns,
- ✓ Growing Patterns,
- ✓ Patterns: Numbers,
- ✓ Finding Rules,
- ✓ Algebraic Word Problems,
- ✓ Evaluating Outputs,

Repeating Patterns

✍Circle the picture that comes next in each picture pattern.

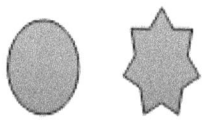

1)

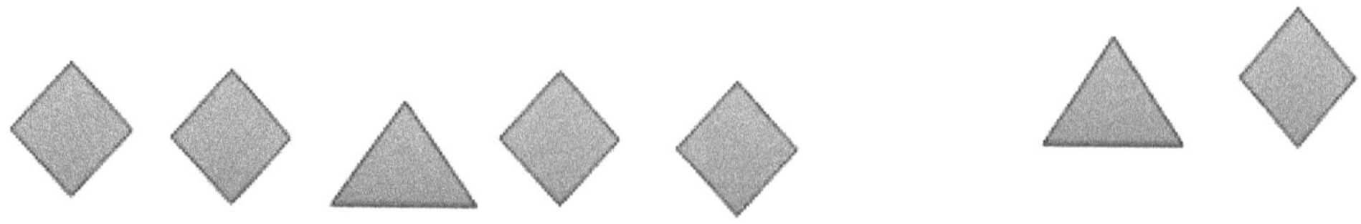

2)

3)

4)

5)

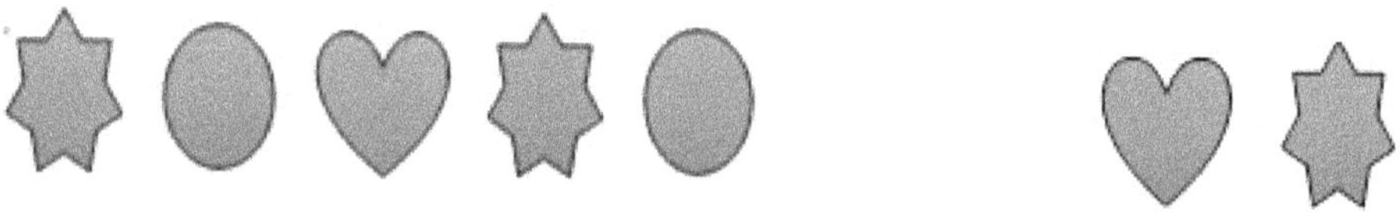

Growing Patterns

Draw the picture that comes next in each growing pattern.

1)

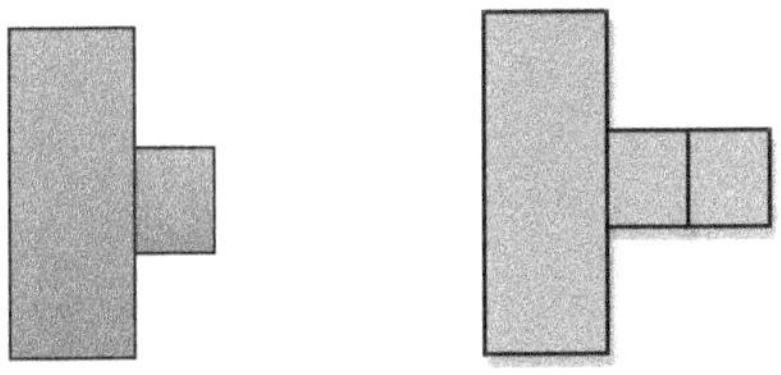

2)

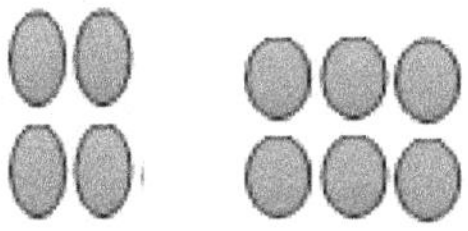

3)

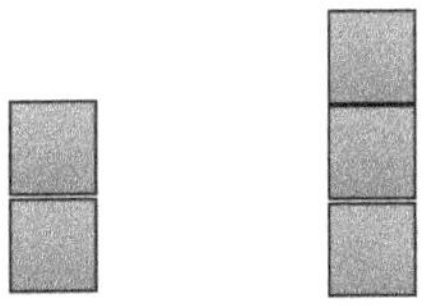

4)

5)

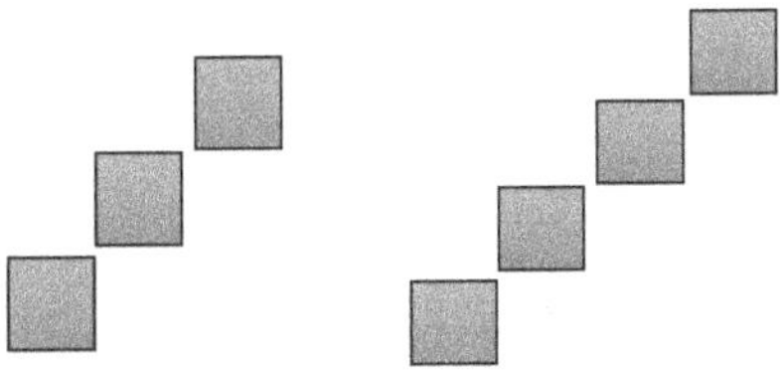

Patterns: Numbers

Write the numbers that come next.

1) 2, 5, 8, 11, ____, ____, ____, ____

2) 10, 15, 20, 25, ____, ____, ____, ____

3) 4, 8, 12, 16, ____, ____, ____, ____

4) 7, 17, 27, 37, ____, ____, ____, ____

5) 5, 12, 19, 26, ____, ____, ____, ____

6) 8, 16, 24, 32, 40, ____, ____, ____, ____

Write the next three numbers in each counting sequence.

7) –31, –19, –7, _____, _____, _____, _____

8) 541, 526, 511, _____, _____, _____, _____

9) 14, 34, _____, _____, 94, _____

10) 21, 29, _____, _____, _____

11) 89, 78, _____, _____, _____

12) 95, 82, 69, _____, _____, _____

13) 198, 166, 134, _____, _____, _____

14) What are the next three numbers in this counting sequence?

1870, 1970, 2070, _____, _____, _____

15) What is the fourth number in this counting sequence?

8, 14, 20, _____

Finding Rules

Complete the output.

1- **Rule:** the output is $x - 10.5$

Input	x	15	18	27	32.25	48.5
Output	y					

1) **Rule:** the output is $x \times 5\frac{1}{3}$

Input	x	3	9	15	21	33
Output	y					

2- **Rule:** the output is $x \div 9$

Input	x	513	387	342	198	126
Output	y					

Find a rule to write an expression.

3- **Rule:** ____________________

Input	x	4	14	19	24
Output	y	10	35	47.5	60

4- **Rule:** ____________________

Input	x	5	13	19.6	34.5
Output	y	14.4	22.4	29	43.9

5- **Rule:** ____________________

Input	x	72	96	132	230.4
Output	y	9	12	16.5	28.8

Algebraic Word Problems

Circle the number sentence that fits the problem. Then solve for x.

1) Mary had $42. Then she earned more money (x). Now she has $86.

$\$42 + x = \$86$ OR $\$42 + \$86 = x$

$x =$ ____

2) Lisa had $35. Then she earned more money (x). Now she has $78.

$\$35 + x = \$78$ OR $\$35 + \$78 = x$

$x =$ ____

3) Matthew had $37. Then he earned more money (x). Now he has $98.

$\$37 + x = \$98$ OR $\$37 + \$98 = x$

$x =$ ____

4) Charlotte gave 19 of the cookies he had baked to a friend and now he has 45 cookies left. $45 - 19 = x$ OR $x - 19 = 45$

$x =$ ____

5) Mia gave 32 of the cookies she had baked to a friend and now she has 55 cookies left. $55 - 32 = x$ OR $x - 32 = 55$

$x =$ ____

6) Lucas gave 41 of the cookies he had baked to a friend and now he has 49 cookies left. . $49 - 41 = x$ OR $x - 41 = 49$

$x =$ ____

Evaluating Outputs

Write output of each algebraic expression.

1) $11 - x\,,\ x = 3$

2) $x + 14,\ x = 4$

3) $8 - 3x,\ x = 1$

4) $3x + \frac{1}{3},\ x = \frac{1}{2}$

5) $2x + 18,\ x = 1.5$

6) $7 - 3x\,,\ x = 1.2$

7) $12 + 2x - 15,\ x = 3.5$

8) $25 - 5x,\ x = 2.2$

9) $\frac{44}{x} - 58,\ x = 0.4$

10) $\frac{x}{3} - 15 + x, x = 12.6$

11) $\frac{x}{7} + 9.5, x = 28.7$

12) $\frac{33}{x} - 5.2 + 2.1x,\ x = 3$

13) $2x - \frac{45}{x} - 12,\ x = 9$

14) $\frac{x}{17} - 1.8\,,\ x = 34$

15) $2\,(12.5x + 8),\ x = 2$

16) $16x + 13x - 27 + 9,$

$x = 1.5$

17) $8.7 - \frac{16}{x} + 3x,$

$x = 4$

18) $5(3a - 2a),$

$a = 5.5$

19) $14 - 2x + 16 - x,$

$x = 3.5$

20) $6x - 3 - x,$

$x = 1.6$

21) $18 - 2(2x + x),\ x = 0.2$

Answers of Worksheets

Repeating pattern

1) 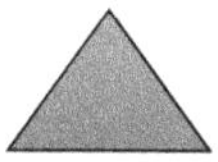

2) 

3) 

4) 

5) 

Growing patterns

1) 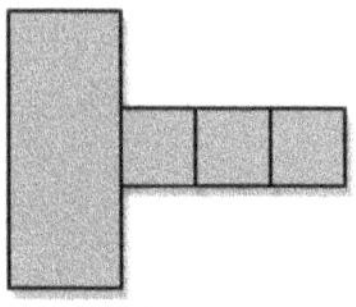

2) 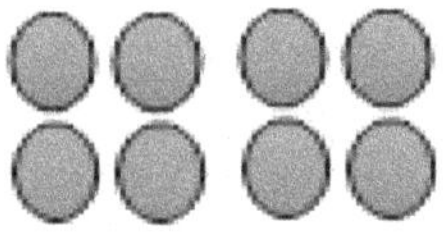

3) 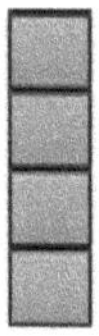

4) 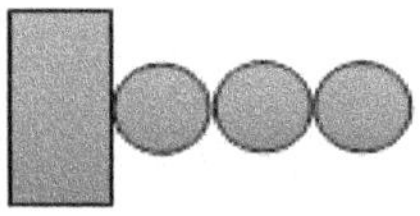

5) 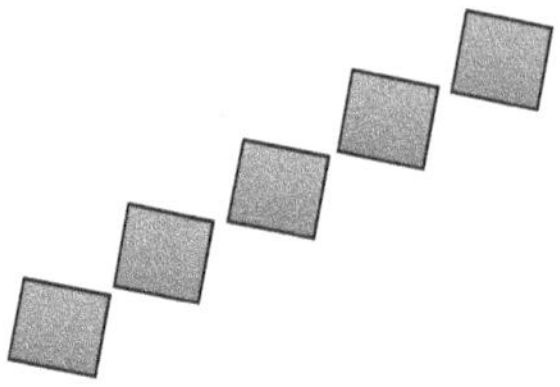

Patterns: Numbers

1) 2, 5, 8, 11, 14, 17, 20, 23
2) 10, 15, 20, 25, 30, 35, 40, 45
3) 4, 8, 12, 16, 20, 24, 28, 32
4) 7, 17, 27, 37, 47, 57, 67, 77
5) 5, 12, 19, 26, 33, 40, 47, 54
6) 8, 16, 24, 32, 40, 48, 56, 64
7) 5, 17, 29, 41
8) 496, 481, 466, 451
9) 14, 34, 54, 74, 94, 114
10) 37, 45, 53
11) 67, 56, 45
12) 56, 43, 30
13) 102,70, 38
14) 2170, 2270, 2370
15) 26

Finding Rules

1)

Input	x	15	18	27	32.25	48.5
Output	y	4.5	7.5	16.5	21.75	38

2)

Input	x	3	9	15	21	33
Output	y	16	48	80	112	176

3)

Input	x	513	387	342	198	126
Output	y	57	43	38	22	14

4) $y = 2.5x$

5) $y = x + 9.4$

6) $y = x \div 8$

Algebraic Word Problems

1) $\$42 + x = \$86;\ x = 44$

2) $\$35 + x = \$78;\ x = 43$

3) $\$37 + x = \$98;\ x = 61$

4) $x - 19 = 45;\ x = 64$

5) $x - 32 = 55;\ x = 87$

6) $x - 41 = 49;\ x = 90$

Evaluating Outputs

1) 8

2) 18

3) 5

4) $\frac{11}{6}$

5) 21

6) 3.4

7) 4

8) 14

9) 52

10) 1.8

11) 13.6

12) 12.1

13) 1

14) 0.2

15) 66

16) 25.5

17) 16.7

18) 27.5

19) 19.5

20) 5

21) 16.8

Chapter 7 : Measurement

Topics that you'll learn in this chapter:

- ✓ Reference Measurement Units,
- ✓ Metric Length Units,
- ✓ Customary Length Units,
- ✓ Metric Capacity Units,
- ✓ Customary Capacity Units,
- ✓ Metric Weight and Mass Units,
- ✓ Customary Weight and Mass Units,
- ✓ Temperature Units,
- ✓ Time,
- ✓ Add Money Amounts,
- ✓ Subtract Money Amounts,
- ✓ Money: Word Problems,

Reference Measurement Units

LENGTH

Customary	Metric
1 mile (mi) = 1,760 yards (yd)	1 kilometer (km) = 1,000 meters (m)
1 yard (yd) = 3 feet (ft)	1 meter (m) = 100 centimeters (cm)
1 foot (ft) = 12 inches (in.)	1 centimeter(cm)= 10 millimeters(mm)

VOLUME AND CAPACITY

Customary	Metric
1 gallon (gal) = 4 quarts (qt)	1 liter (L) = 1,000 milliliters (mL)
1 quart (qt) = 2 pints (pt.)	
1 pint (pt.) = 2 cups (c)	
1 cup (c) = 8 fluid ounces (Fl oz)	

WEIGHT AND MASS

Customary	Metric
1 ton (T) = 2,000 pounds (lb.)	1 kilogram (kg) = 1,000 grams (g)
1 pound (lb.) = 16 ounces (oz)	1 gram (g) = 1,000 milligrams (mg)

Time

1 year = 12 months

1 year = 52 weeks

1 week = 7 days

1 day = 24 hours

1 hour = 60 minutes

1 minute = 60 seconds

Metric Length Units

Convert to the units.

1) 300 mm = _______ cm

2) 8 m = _________ mm

3) 4.5 m = _________ cm

4) 7 km = _________ m

5) 9,400mm = ______ m

6) 1,100 cm = ________ m

7) 2.8 m = _________ cm

8) 4,000 mm = ______ cm

9) 7,000 mm = ______ m

10) 2 km = ________ mm

11) 14.9 km = ________ m

12) 20 m = ________ cm

13) 5,000 m = _______ km

14) 7,600 m = _______ km

Customary Length Units

Convert to the units.

1) 8 ft = _____ in

2) 4 ft = _____ in

3) 6 yd = _____ ft

4) 10 yd = _____ ft

5) 3,520 yd = _____ mi

6) 60 in = _____ ft

7) 144 in = ____ yd

8) 0.5 mi = ______ yd

9) 15 yd = ______ in

10) 42 yd = ______ in

11) 99 ft = ______ yd

12) 1.5 mi = ______ yd

13) 84 in = ______ ft

14) 30 yd = ______ feet

Metric Capacity Units

Convert the following measurements.

1) 32.4 l =__________ ml

2) 7.1 l = __________ ml

3) 54 l = __________ ml

4) 92 l = __________ ml

5) 48 l = __________ ml

6) 13 l = __________ ml

7) 750 ml = __________ l

8) 2,400 ml = ______ l

9) 73,000 ml = ________ l

10) 8,000 ml = ________ l

11) 49,000 ml = ________ l

12) 5,500 ml =_______ l

Customary Capacity Units

Convert the following measurements.

1) 51gal = ______ qt.

2) 35gal = ______ pt.

3) 68gal = _______ c.

4) 20pt. = _______ c

5) 12.5 qt =________ pt.

6) 22.5qt=________ c

7) 51pt. = _______ c

8) 48c =_________ gal

9) 96pt. = ______ gal

10) 136qt =______ gal

11) 15c =______ fl oz

12) 44c = ________ qt

13) 240c = _______ pt.

14) 148 qt= _______ gal

15) 160pt. = ______ qt

16) 104 fl oz = _______ c.

Metric Weight and Mass Units

Convert.

1) 60 kg = __________ g

2) 24 kg = __________ g

3) 610 kg = __________ g

4) 82 kg = __________ g

5) 95.8 kg = __________ g

6) 4.85 kg = __________ g

7) 1.5 kg = __________ g

8) 95,000 g = __________ kg

9) 241,000 g = __________ kg

10) 700,000 g = ________ kg

11) 2,500 g = ________ kg

12) 28,900 g = __________kg

13) 970,000 g = ________ kg

14) 325,500 g = ________ kg

Customary Weight and Mass Units

Convert.

1) 10,000 lb. = ______ T

2) 19,000 lb. = _____ T

3) 32,000 lb. = ______ T

4) 16,800lb. = _____ T

5) 27 lb. =_________ oz

6) 25.4 lb.=_________ oz

7) 124 lb. = _________ oz

8) 4T =______ lb.

9) 7T = ______ lb.

10) 11.2T =_____ lb.

11) 12.8T =______ lb.

12) $\frac{6}{5}$ T = _______ oz

13) 9.125 T = _______ oz

14) $\frac{3}{4}$ T=_______oz

Time

Convert to the units.

1) 22 hr. = __________ min

2) 12.5 year = ________ week

3) 5.2hr = ___________ sec

4) 21min = _________ sec

5) 1,800min = _______ hr.

6) 730 day = ________ year

7) 1.5year = ___________ hr.

8) 42 day = _________ hr.

9) 2 day = _________ min

10) 660 min = ________ hr.

11) 10year = ______ month

12) 2,124sec = _______min

13) 168 hr = _________ day

14) 19 weeks = _______ day

How much time has passed?

1) From 2:25 A.M. to 5:35 A.M.: ____ hours and ___ minutes.

2) From 3:30 A.M. to 7:55 A.M.: ____ hours and ___ minutes.

3) It's 6:30 P.M. What time was 2 hours ago? _______ O'clock

4) 4:30 A.M to 7:50 AM: _____ hours and _____ minutes.

5) 4:15 A.M to 8:35 AM: _____ hours and _____ minutes.

6) 5:10 A.M. to 7:35 AM. = _____ hour(s) and _____ minutes.

7) 10:55 A.M. to 3:25 PM. = _____ hour(s) and _____ minutes

8) 8:05 A.M. to 8:40 A.M. = _______ minutes

9) 6:02 A.M. to 6:49 A.M. = _______ minutes

Money Amounts

Add.

1) $\begin{array}{r} \$128 \\ +\$328 \\ \hline \end{array}$ $\begin{array}{r} \$225 \\ +\$245 \\ \hline \end{array}$ $\begin{array}{r} \$150 \\ +\$186 \\ \hline \end{array}$

2) $\begin{array}{r} \$453 \\ +\$128 \\ \hline \end{array}$ $\begin{array}{r} \$440 \\ +\$541 \\ \hline \end{array}$ $\begin{array}{r} \$258 \\ +\$248 \\ \hline \end{array}$

3) $\begin{array}{r} \$645 \\ +\$112.5 \\ \hline \end{array}$ $\begin{array}{r} \$235.4 \\ +\$452.1 \\ \hline \end{array}$ $\begin{array}{r} \$125.99 \\ +\$148.32 \\ \hline \end{array}$

4) $\begin{array}{r} \$321.40 \\ +\$175.80 \\ \hline \end{array}$ $\begin{array}{r} \$458.10 \\ +\$752.65 \\ \hline \end{array}$ $\begin{array}{r} \$652.00 \\ +\$324.70 \\ \hline \end{array}$

Subtract.

5) $\begin{array}{r} \$725 \\ -\$334 \\ \hline \end{array}$ $\begin{array}{r} \$543 \\ -\$248 \\ \hline \end{array}$ $\begin{array}{r} \$349 \\ -\$122 \\ \hline \end{array}$

6) $\begin{array}{r} \$658.20 \\ -\$220.30 \\ \hline \end{array}$ $\begin{array}{r} \$752.10 \\ -\$452.15 \\ \hline \end{array}$ $\begin{array}{r} \$312.50 \\ -\$89.90 \\ \hline \end{array}$

7) $\begin{array}{r} \$315.90 \\ -\$220.10 \\ \hline \end{array}$ $\begin{array}{r} \$548.40 \\ -\$342.10 \\ \hline \end{array}$ $\begin{array}{r} \$968.40 \\ -\$324.50 \\ \hline \end{array}$

8) Linda had $18.60. She bought some game tickets for $9.25. How much did she have left?

Money: Word Problems

Solve.

1) How many boxes of envelopes can you buy with $40 if one box costs $8?

2) After paying $7.22 for a salad, Ella has $45.86. How much money did she have before buying the salad?

3) How many packages of diapers can you buy with $96 if one package costs $6?

4) Last week James ran 28.5 miles more than Michael. James ran 59 miles. How many miles did Michael run?

5) Last Friday Jacob had $14.68. Over the weekend he received some money for cleaning the attic. He now has $38.95. How much money did he receive?

6) After paying $3.15 for a sandwich, Amelia has $48.69. How much money did she have before buying the sandwich?

Answers of Worksheets

Metric length

1) 30 cm
2) 8,000 mm
3) 450 cm
4) 7,000 m
5) 9.4 m
6) 11 m
7) 280 cm
8) 400 cm
9) 7 m
10) 2,000,000 mm
11) 14,900 m
12) 2,000 cm
13) 5 km
14) 7.6 km

Customary Length

1) 96
2) 48
3) 18
4) 30
5) 2
6) 5
7) 4
8) 880
9) 540
10) 1,512
11) 33
12) 2,640
13) 7
14) 90

Metric Capacity

1) 32,400 ml
2) 7,100 ml
3) 54,000 ml
4) 92,000 ml
5) 48,000 ml
6) 13,000 ml
7) 0.75ml
8) 2.4 ml
9) 73 ml
10) 8L
11) 49 L
12) 5.5 L

Customary Capacity

1) 204 qt
2) 280 pt.
3) 1,088 c
4) 40 c
5) 25 pt.
6) 90c
7) 102 c
8) 3 gal
9) 12 gal
10) 34 gal
11) 120 qt
12) 11qt
13) 120 pt.
14) 37 gal
15) 80 qt
16) 13 pt.

Metric Weight and Mass

1) 60,000 g
2) 24,000 g
3) 610,000 g
4) 82,000 g
5) 95,800g
6) 4,850 g
7) 1,500 g
8) 95 kg
9) 241 kg
10) 700 kg
11) 2.5 kg
12) 28.9 kg
13) 970 kg
14) 325.5 kg

Customary Weight and Mass

1) 5 T
2) 9.5 T
3) 16 T
4) 8.4 T
5) 432 oz
6) 406.4 oz

7) 1,984 oz
8) 8,000 lb.
9) 14,000 lb.
10) 22,400 lb.
11) 25,600 lb.
12) 38,400 oz
13) 292,000 oz
14) 24,000 oz

Time - Convert

1) 1,320 min
2) 650 weeks
3) 18,720 sec
4) 1,260 sec
5) 30 hr
6) 2 year
7) 13,140 hr
8) 1,008 hr
9) 2,880 min
10) 11hr
11) 120 months
12) 35.4 min
13) 7 days
14) 133 days

Time - Gap

1) 3:10
2) 4:25
3) 4:30P.M.
4) 3:20
5) 4:20
6) 2:25
7) 4:30
8) 35 minutes
9) 47 minutes

Add Money

1) 456, 470, 336
2) 581, 981, 506
3) 757.5, 687.5, 274.31
4) 497.2, 1210.75, 976.7

Subtract Money

5) 391, 295, 227
6) 437.9, 299.95, 222.6
7) 95.8, 206.3, 643.9
8) $9.35

Money: word problem

1) 5
2) $53.08
3) 16
4) 30.5
5) 24.27
6) 51.84

Chapter 8 : Geometric

Topics that you'll learn in this chapter:

- ✓ Identifying Angles,
- ✓ Estimate and Measure Angles,
- ✓ Polygon Names,
- ✓ Classify Triangles,
- ✓ Parallel Sides in Quadrilaterals,
- ✓ Identify Parallelograms,
- ✓ Identify Trapezoids,
- ✓ Identify Rectangles,
- ✓ Perimeter and Area of Squares,
- ✓ Perimeter and Area of rectangles,
- ✓ Area and Perimeter: Word Problems,
- ✓ Volume,

Identifying Angles

Write the name of the angles (Acute, Right, Obtuse, and Straight).

1)

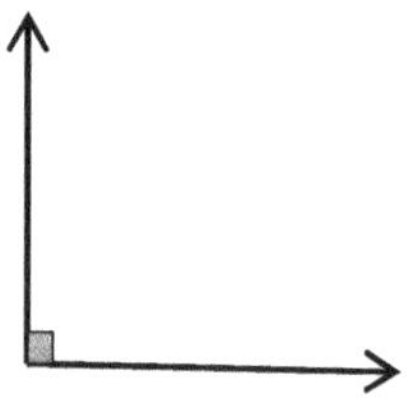

2)

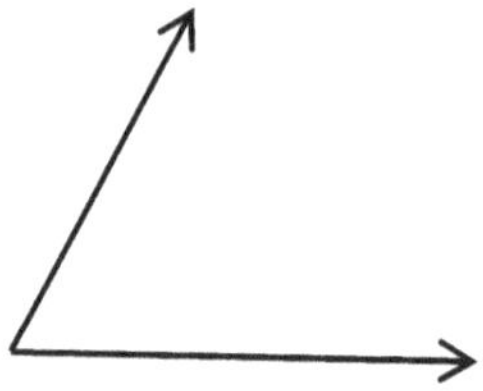

3)

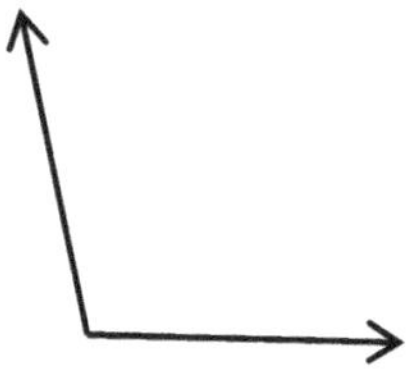

4)

5)

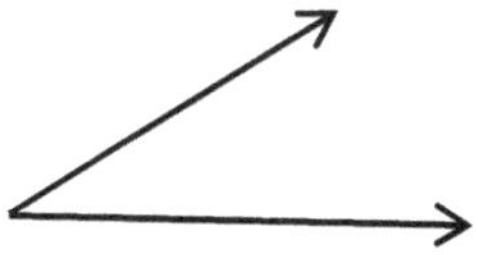

6)

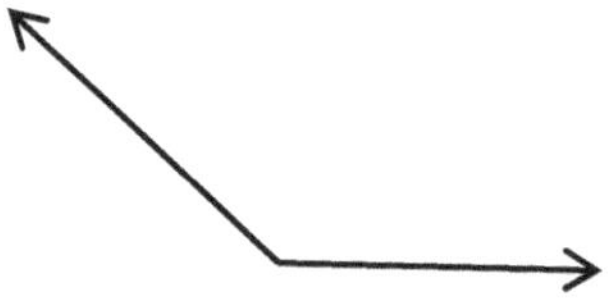

7)

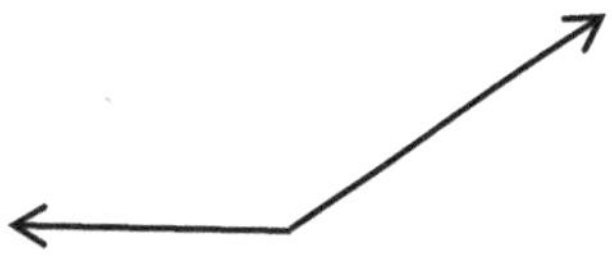

8)

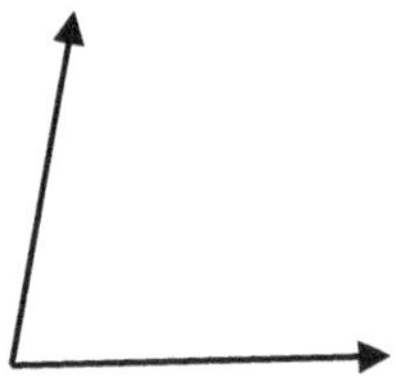

Estimate Angle Measurements

✍Estimate the approximate measurement of each angle in degrees.

1)

2)

3)

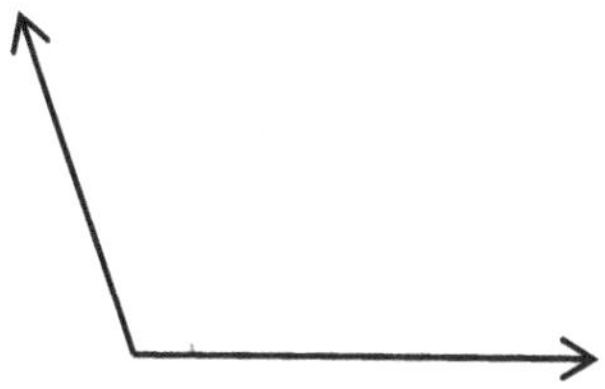

4)

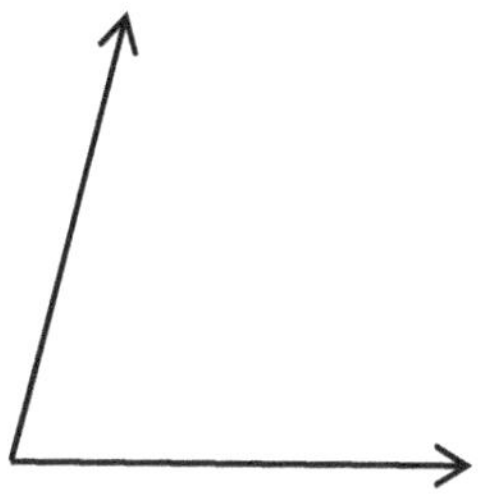

5)

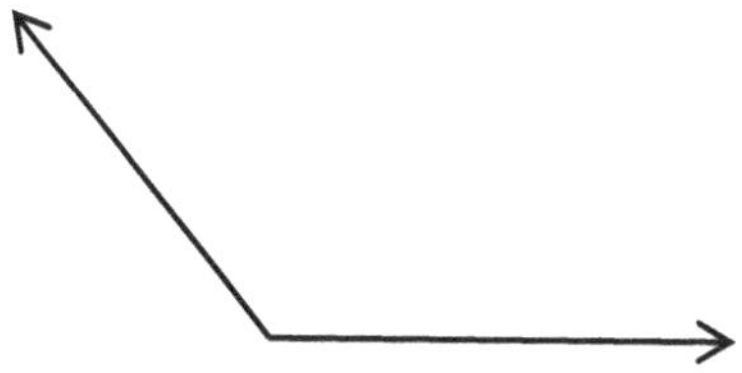

6)

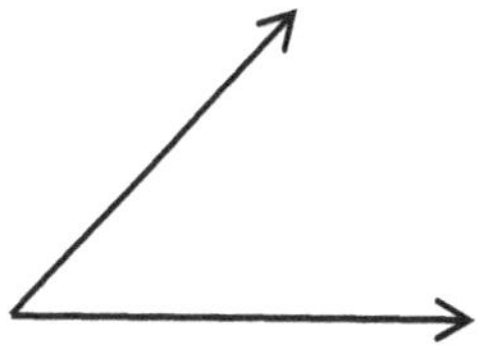

7)

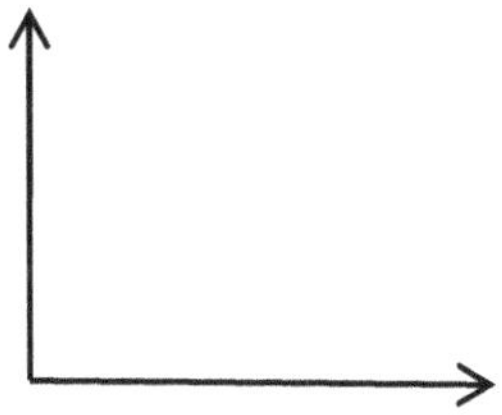

8)

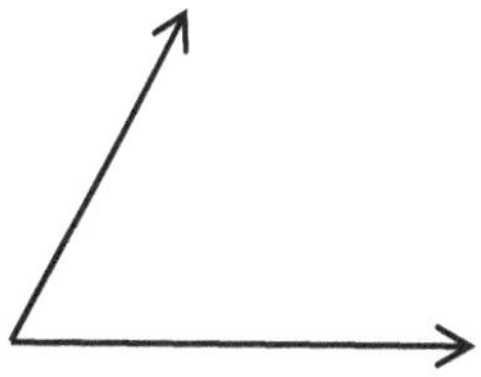

Measure Angles with a Protractor

Use protractor to measure the angles below.

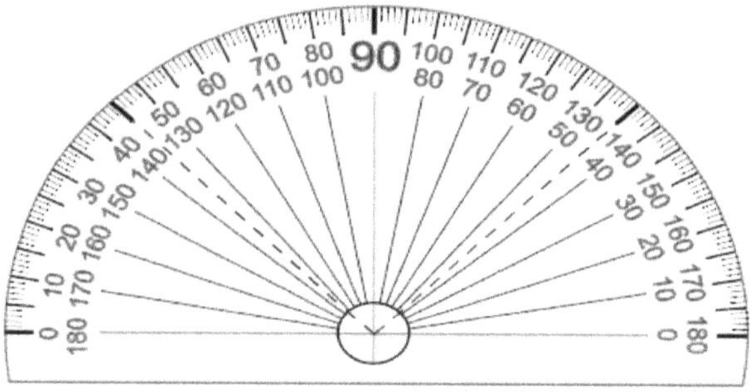

1)

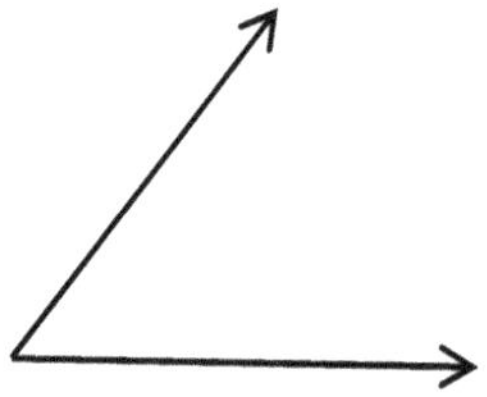

2)

3)

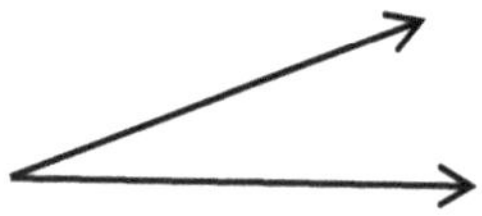

4)

Use a protractor to draw angles for each measurement given.

1) 140°

2) 100°

3) 110°

4) 120°

5) 55°

Polygon Names

Write name of polygons.

1)

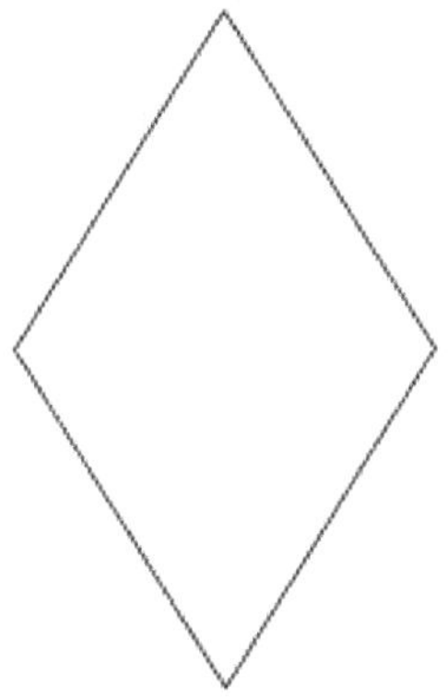

2)

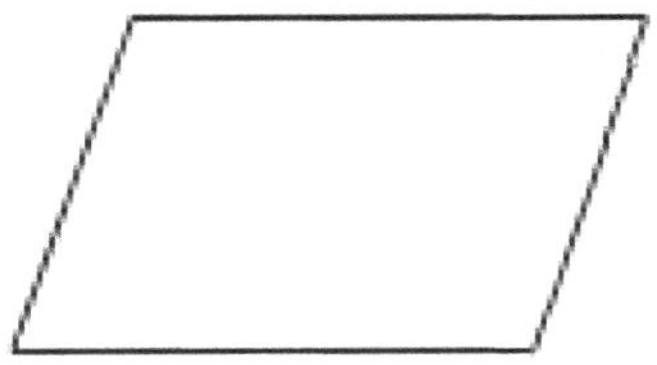

3)

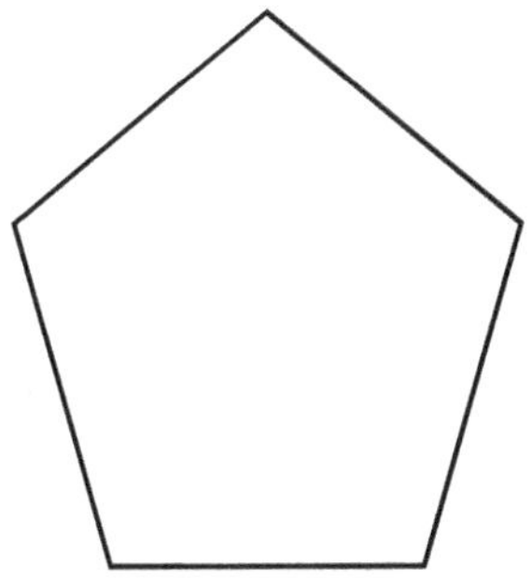

4)

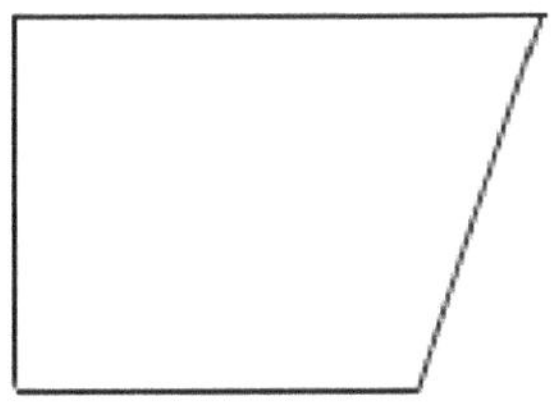

5)

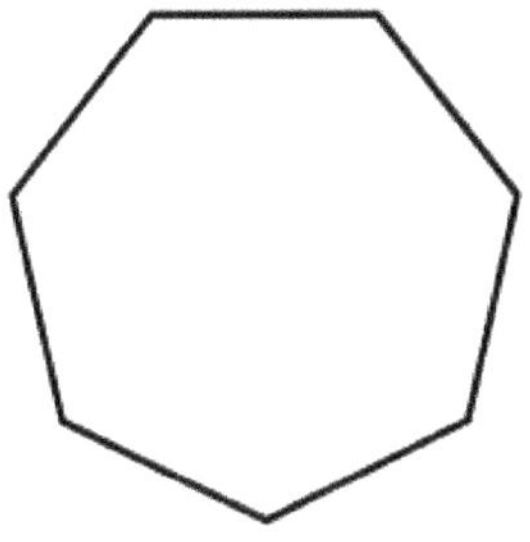

6)

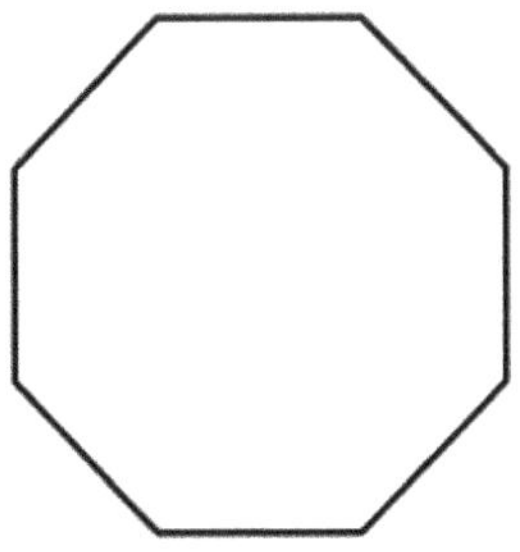

Classify Triangles

✍Classify the triangles by their sides and angles.

1)

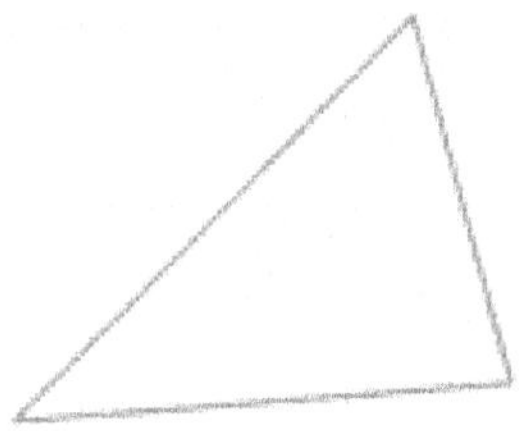

2)

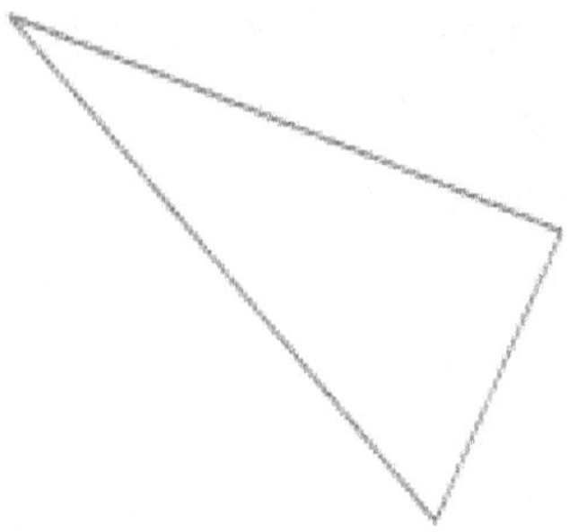

3)

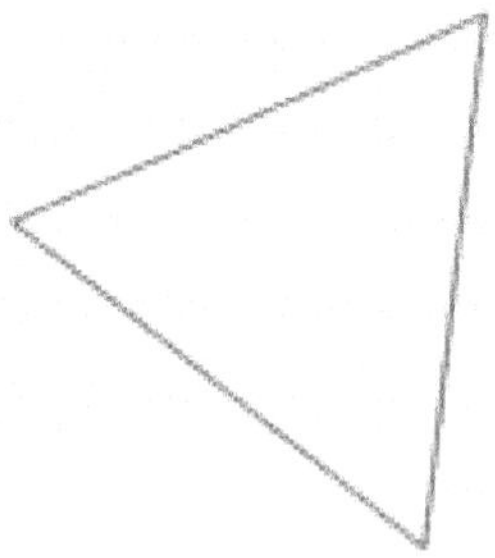

4)

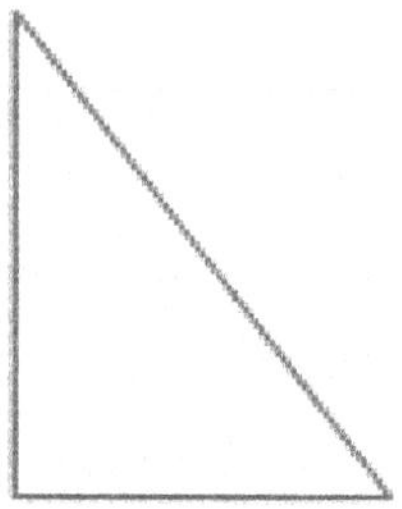

5)

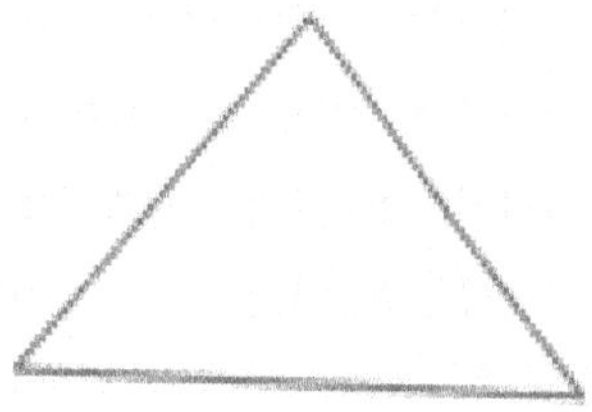

6)

Parallel Sides in Quadrilaterals

Write name of quadrilaterals.

1)

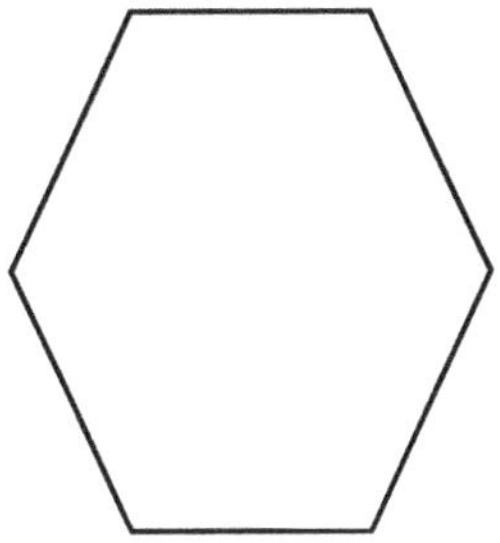

2)

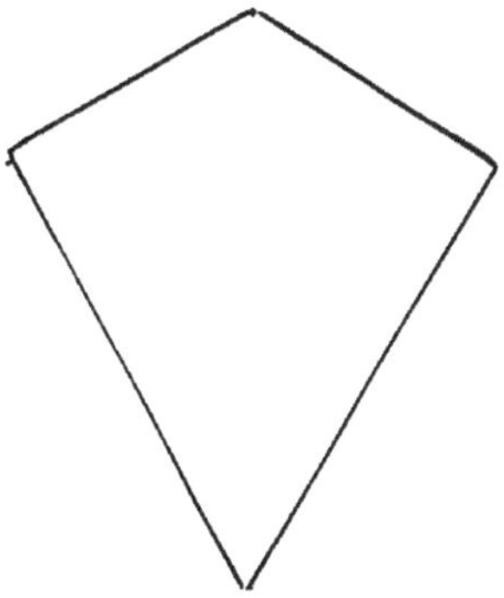

3)

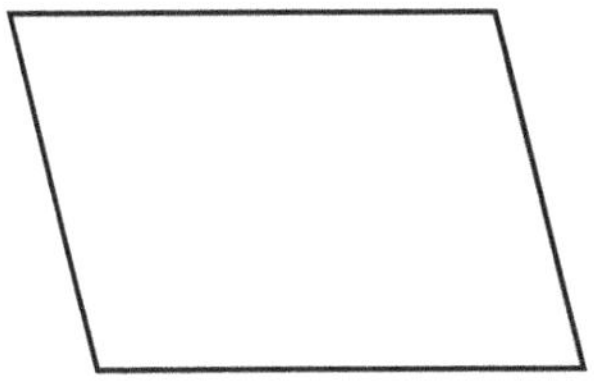

4)

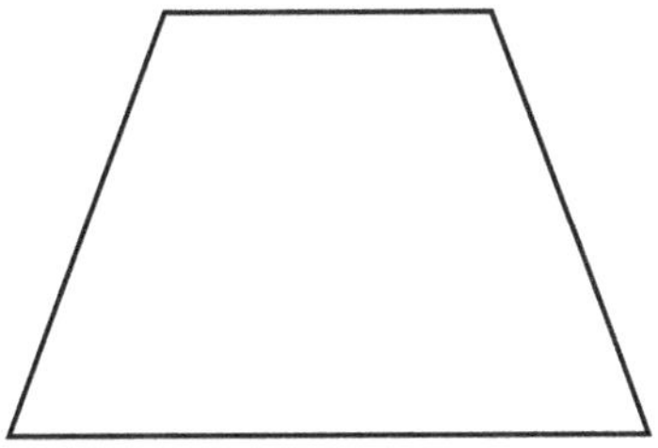

5)

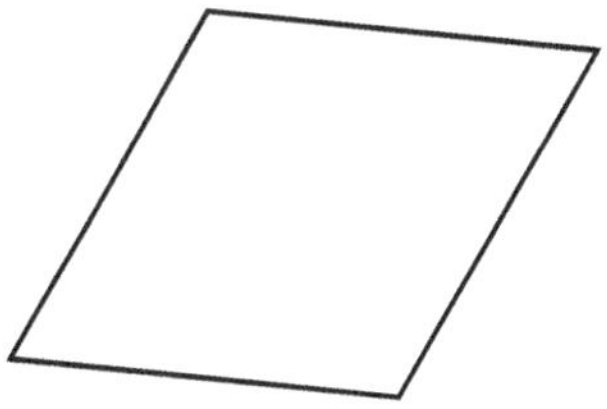

6)

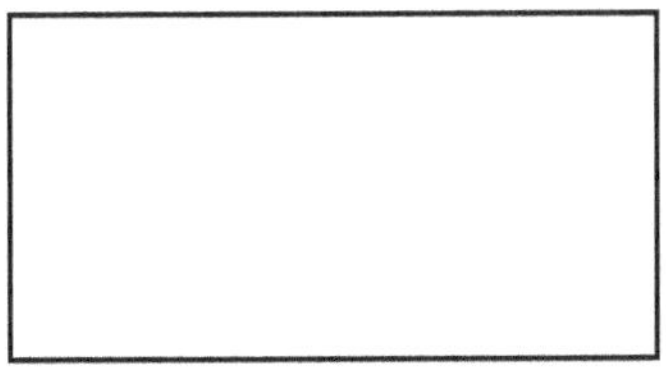

Identify Rectangles

✍Solve.

1) A rectangle has _____ sides and _____ angles.

2) Draw a rectangle that is 5.5centimeters long and 2.5 centimeters wide. What is the perimeter?

3) Draw a rectangle 3.5 cm long and 1.5 cm wide.

4) Draw a rectangle whose length is 4.25 cm and whose width is 2.45 cm. What is the perimeter of the rectangle?

5) What is the perimeter of the rectangle?

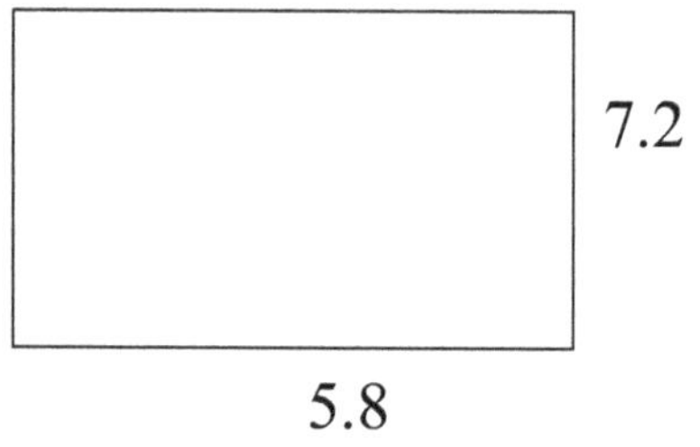

Perimeter: Find the Missing Side Lengths

Find the missing side of each shape.

1) perimeter = 57.2

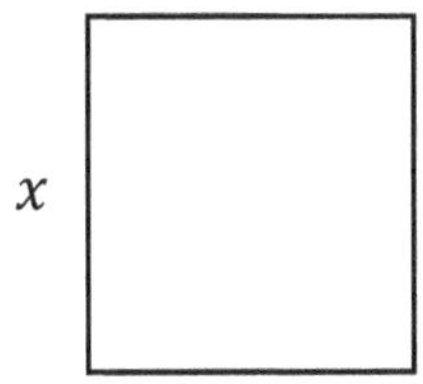

2) perimeter = 21.2

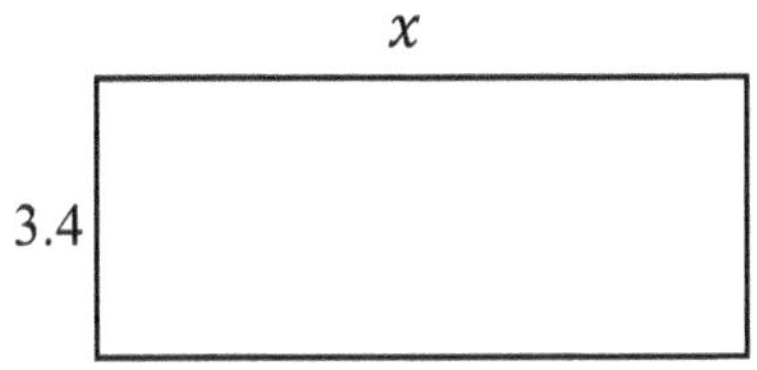

3) perimeter = 27.5

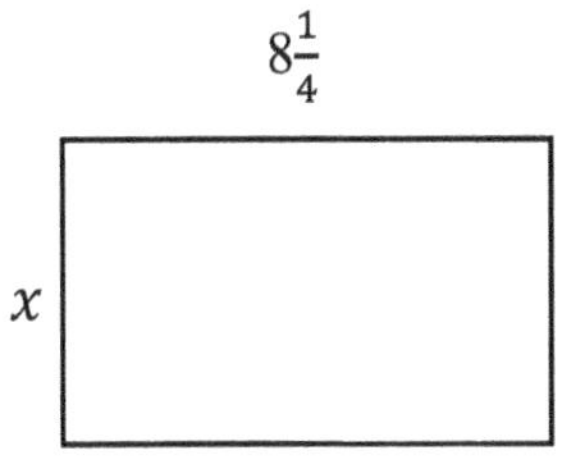

4) perimeter = 35.2

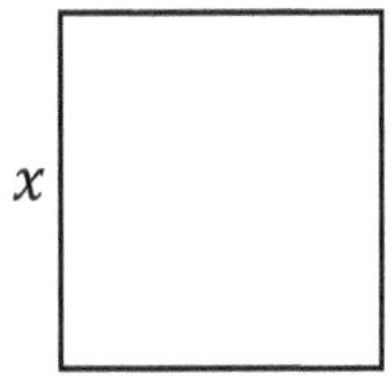

5) perimeter = 75.6

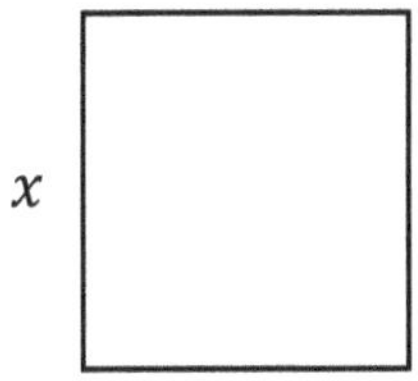

6) perimeter = 30.8

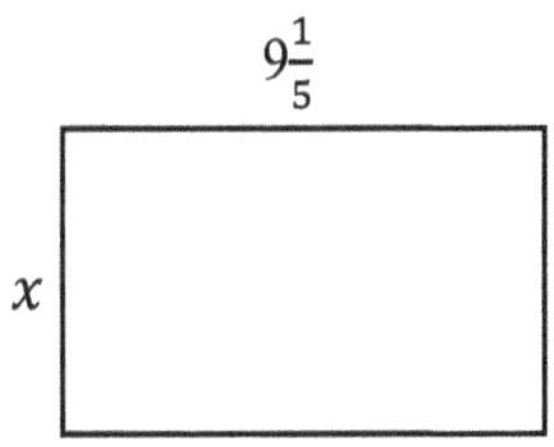

7) perimeter = 36.25

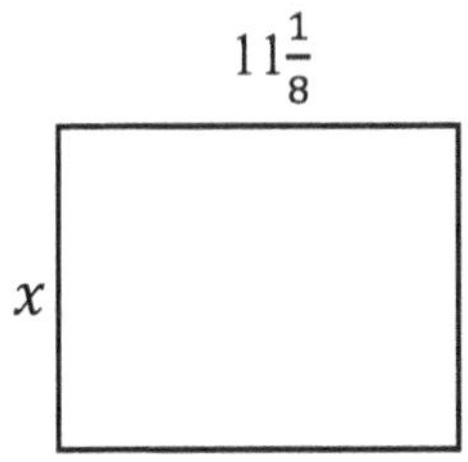

8) perimeter = 46.8

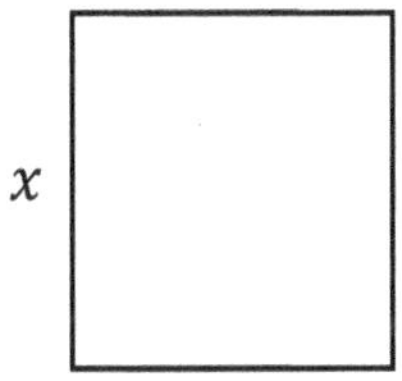

Perimeter and Area of Squares

Find perimeter and area of squares.

1) A: ____, P: ____

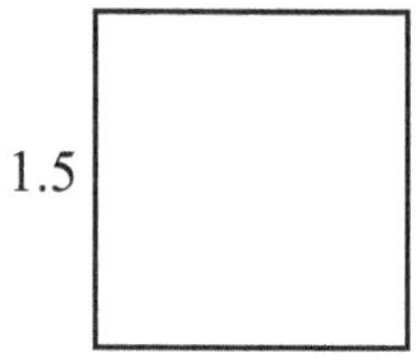

2) A: ____, P: ____

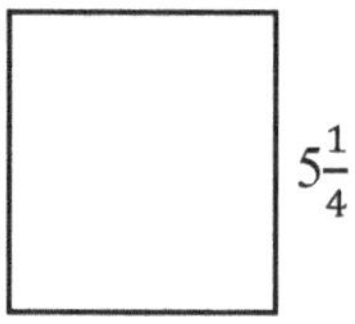

3) A: ____, P: ____

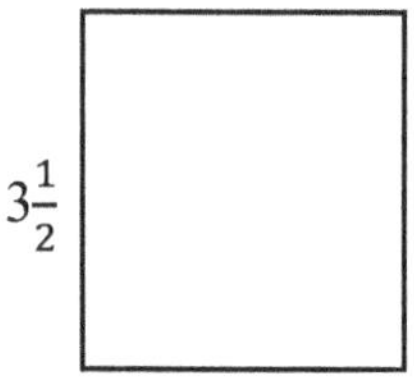

4) A: ____, P: ____

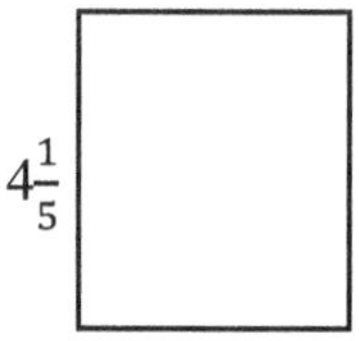

5) A: ____, P: ____

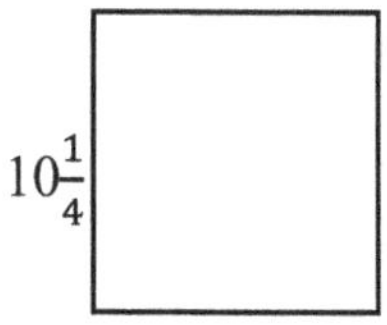

6) A: ____, P: ____

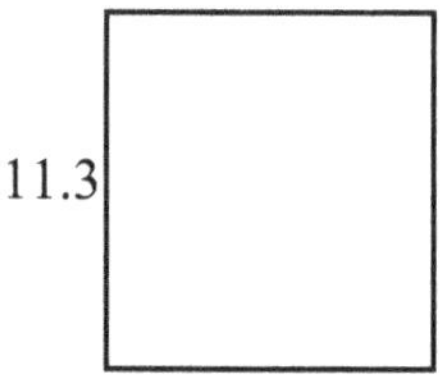

7) A: ____, P: ____

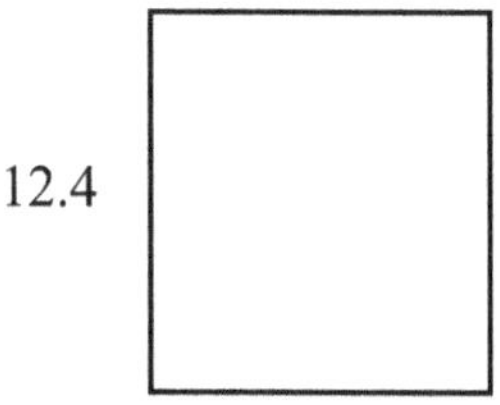

8) A: ____, P: ____

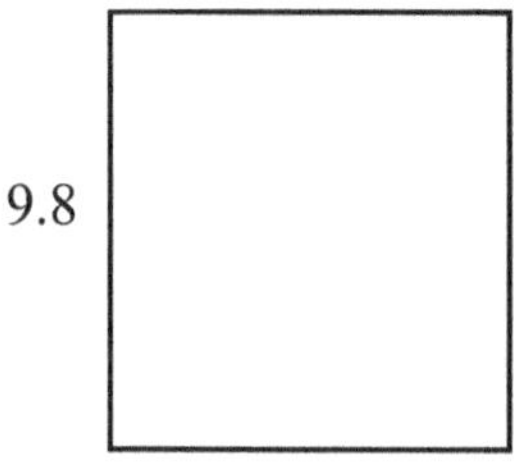

Perimeter and Area of rectangles

Find perimeter and area of rectangles.

1) A: ____, P: ____

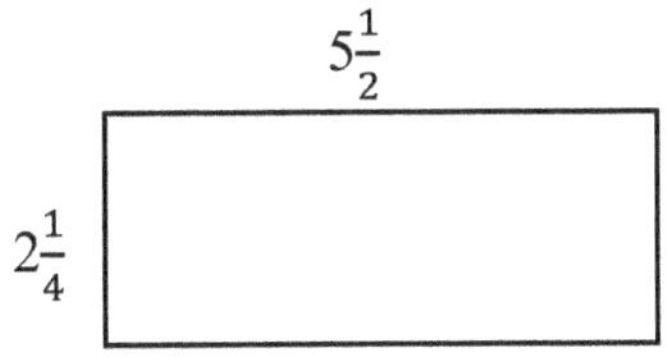

2) A: ____, P: ____

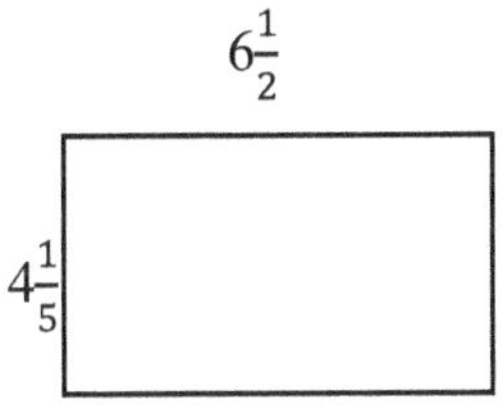

3) A: ____, P: ____

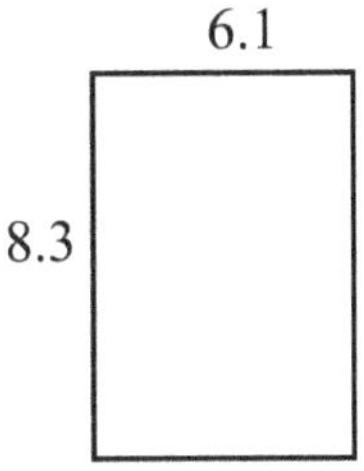

4) A: ____, P: ____

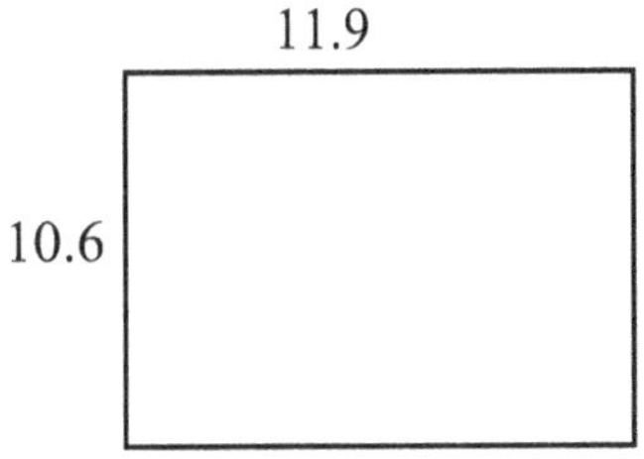

5) A: ____, P: ____

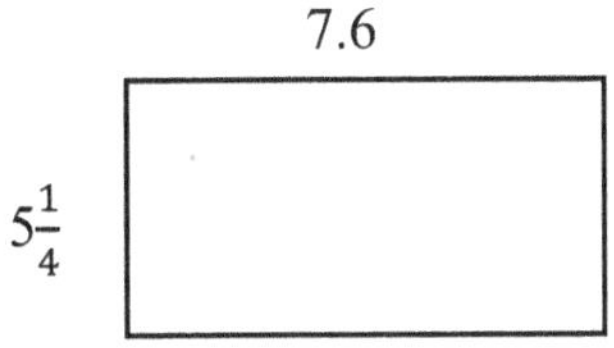

6) A: ____, P: ____

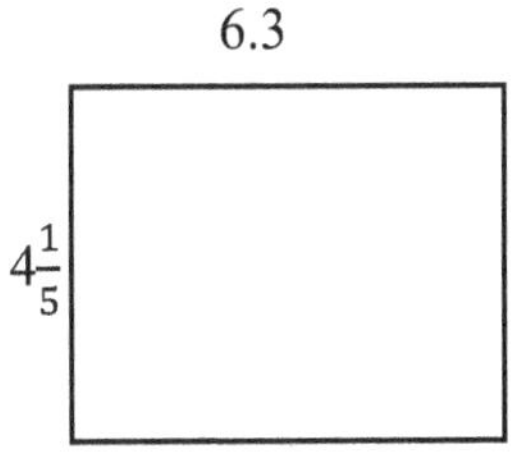

7) A: ____, P: ____

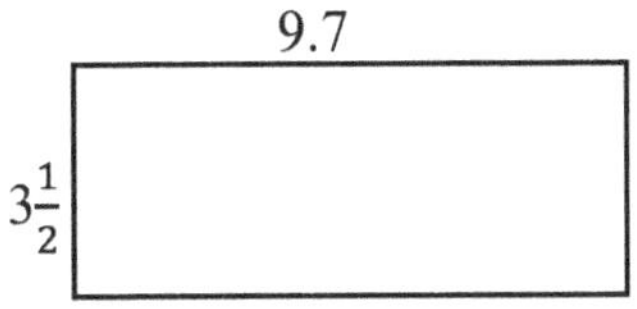

8) A: ____, P: ____

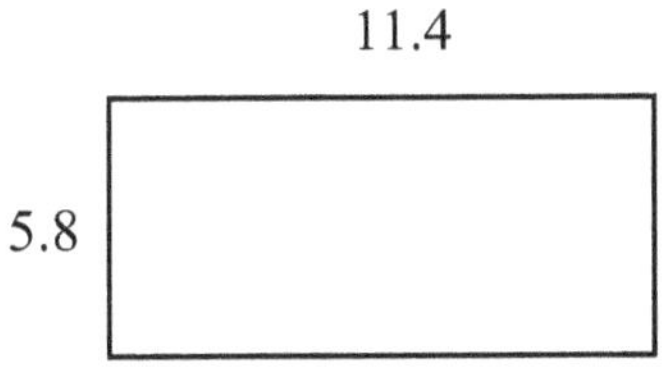

Find the Area or Missing Side Length of a Rectangle

Find area or missing side length of rectangles.

1) Area =?

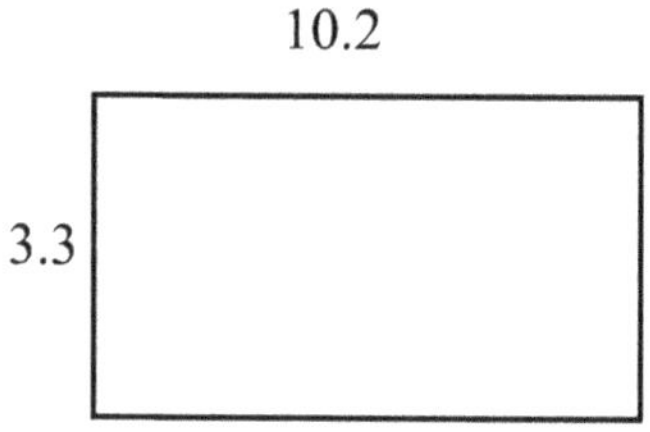

2) Area = 42.12, x=?

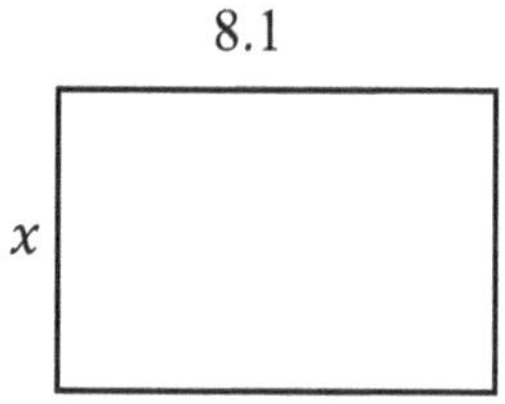

3) Area = 29.52, x=?

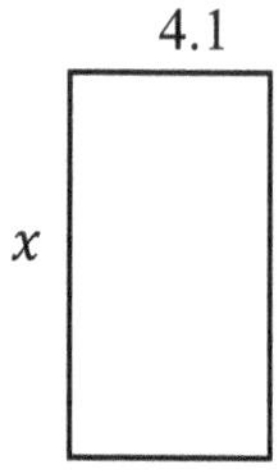

4) Area =?

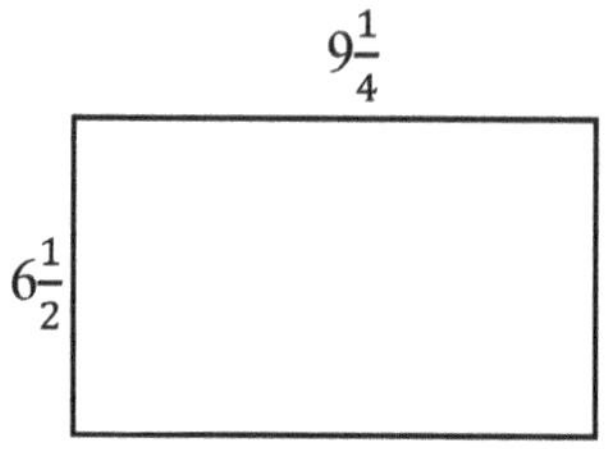

5) Area =?

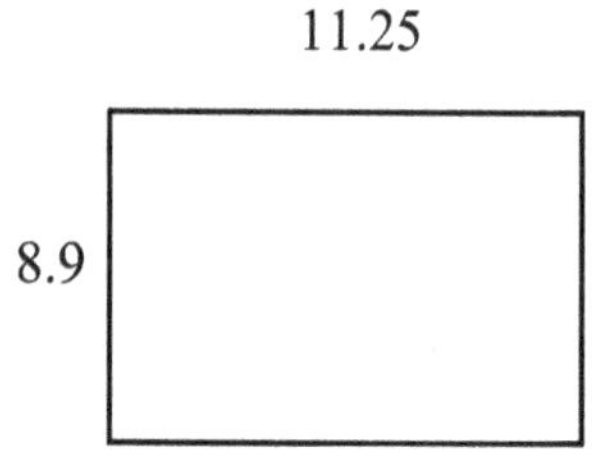

6) Area = 662.34 x=?

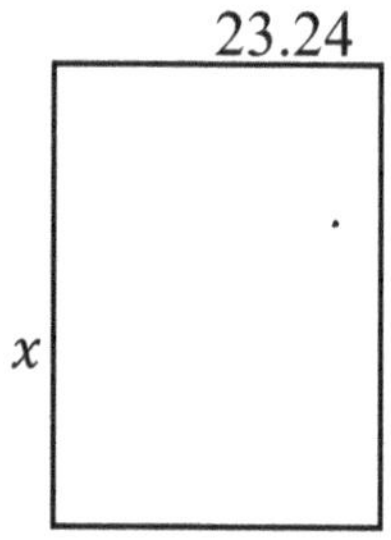

7) Area = 216.24, x=?

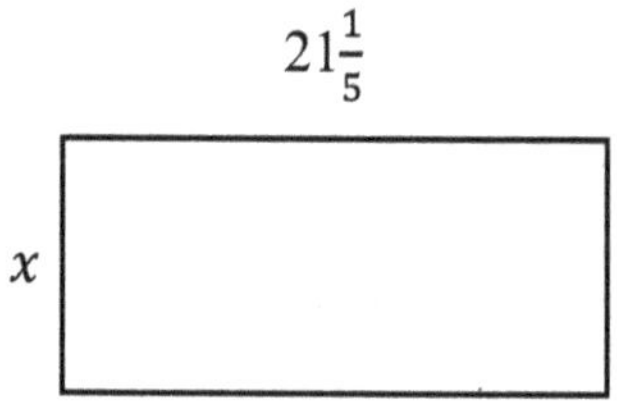

8) Area 336.42, x=?

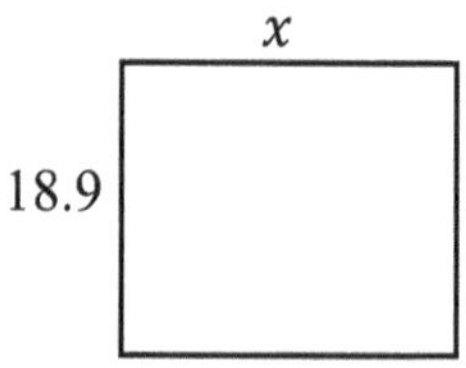

Area and Perimeter: Word Problems

Solve.

1) The area of a rectangle is 90.86 square meters. The width is 7.7 meters. What is the length of the rectangle?

2) A square has an area of 6.25 square feet. What is the perimeter of the square?

3) Ava built a rectangular vegetable garden that is 3.2 feet long and has an area of 21.12 square feet. What is the perimeter of Ava's vegetable garden?

4) A square has a perimeter of 12.8 millimeters. What is the area of the square?

5) The perimeter of David's square backyard is 0.96 meters. What is the area of David's backyard?

6) The area of a rectangle is 37.63 square inches. The length is 7.1 inches. What is the perimeter of the rectangle?

Volume of Cubes and Rectangle Prisms

Find the volume of each of the rectangular prisms.

1)

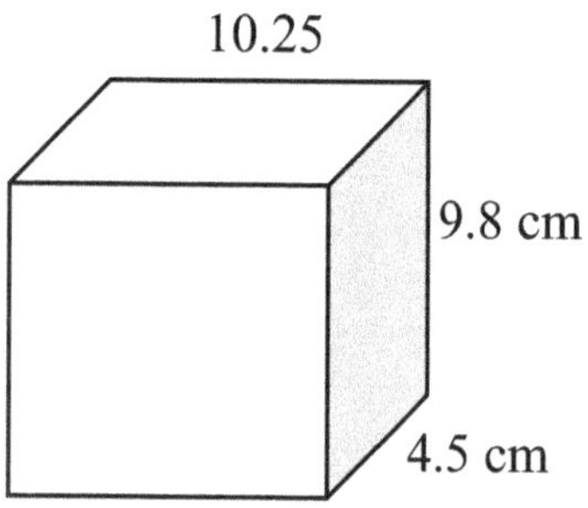

2)

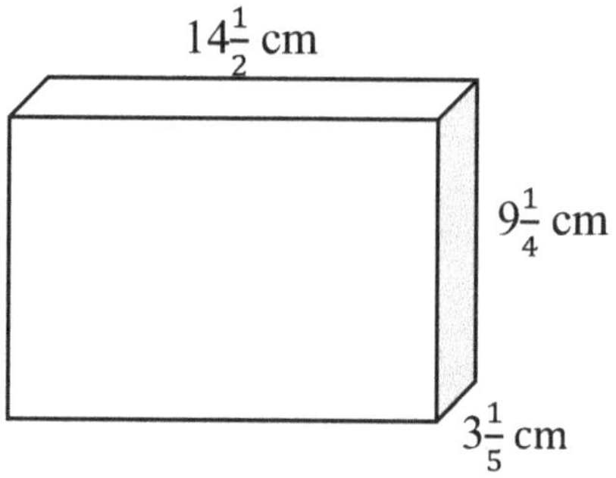

3)

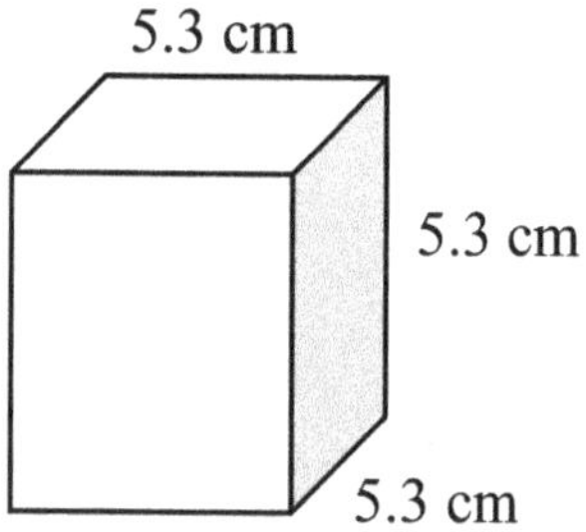

4)

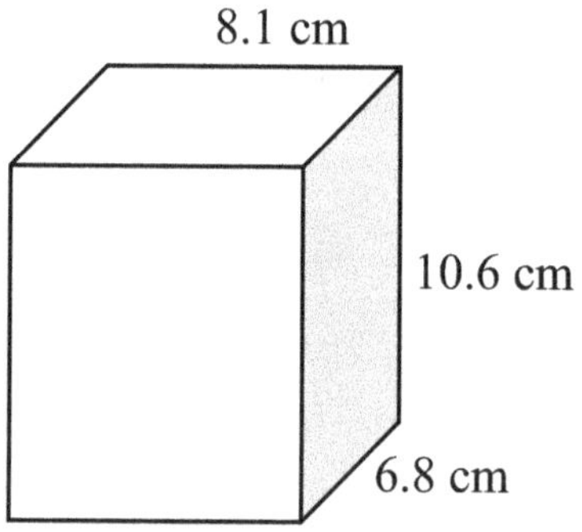

5)

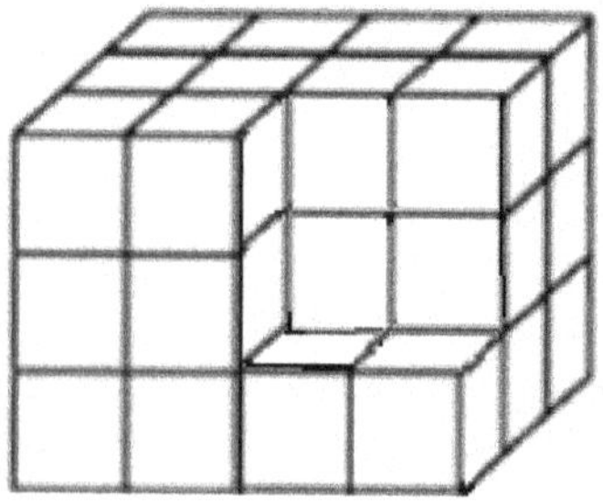

6)

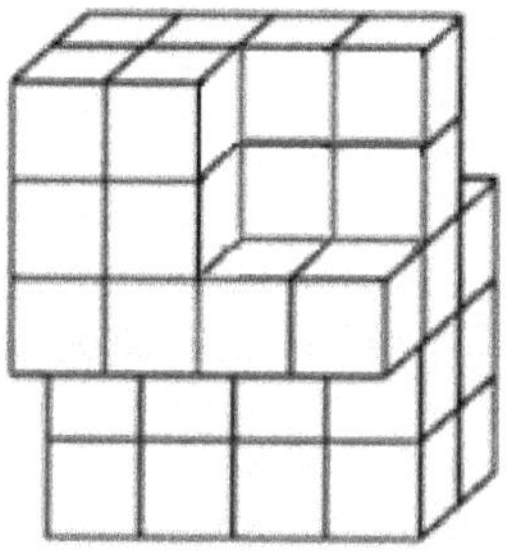

Answers of Worksheets

Identifying Angles

1) Right
2) Acute
3) Obtuse
4) Straight
5) Acute
6) Obtuse
7) Obtuse
8) Acute

Estimate Angle Measurements

1) 160°
2) 180°
3) 110°
4) 75°
5) 130°
6) 45°
7) 90°
8) 60°

Measure Angles with a Protractor

1) 50∘
2) 135∘
3) 20∘
4) 170∘

Draw Angles

1)

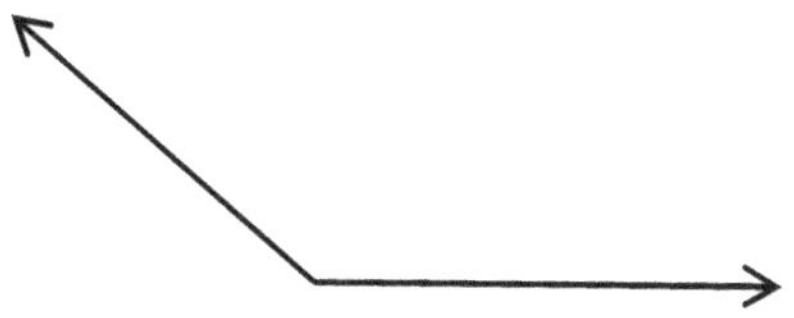

2)

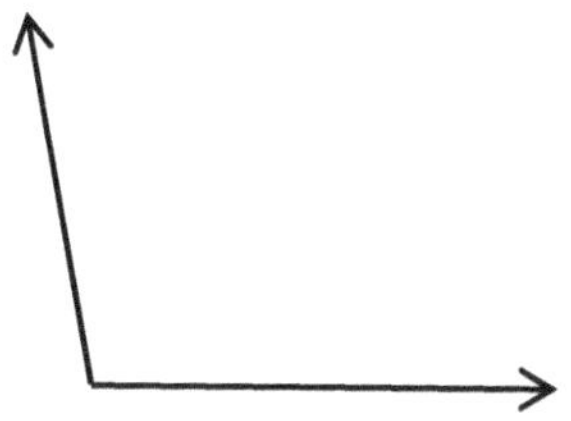

3)

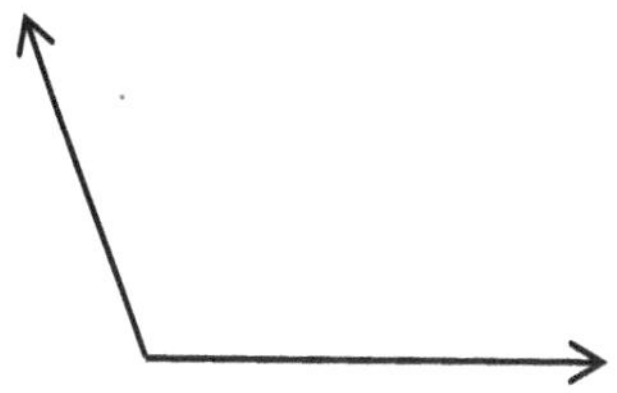

4)

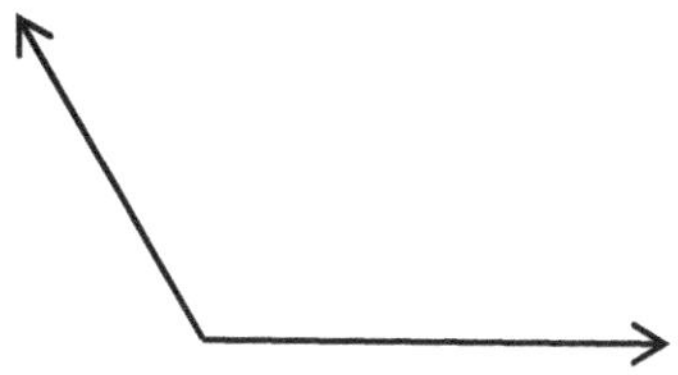

5)

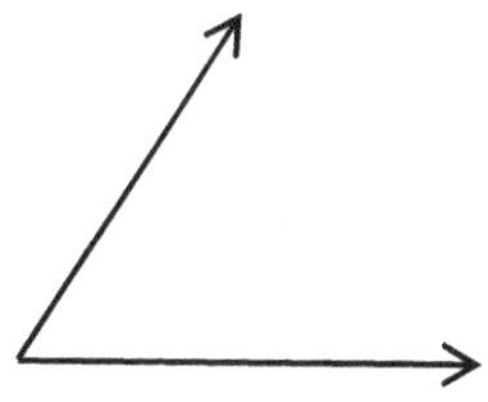

Polygon Names

1) Diamond
2) Parallelogram
3) Pentagon
4) Trapezius
5) Heptagon
6) Octagon

Classify Triangles

1) Scalene, acute
2) Isosceles, acute
3) Equilateral, acute
4) Scalene, right
5) Isosceles, right
6) Scalene, obtuse

Parallel Sides in Quadrilaterals

1) Hexagon
2) Kike
3) Parallelogram
4) Trapezoid
5) Rhombus
6) Rectangle

Identify Rectangles

1) 4 - 4
2) 16
3) Draw the rectangle.
4) 13.4
5) 26

Perimeter: Find the Missing Side Lengths

1) 14.3
2) 7.2
3) 5.5
4) 8.8
5) 18.9
6) 6.2
7) 7
8) 11.7

Perimeter and Area of Squares

1) A: 2.25, P: 6
2) A: 27.56, P: 21
3) A: 12.25, P: 14
4) A: 17.64, P: 16.8
5) A: 105.063 P: 41
6) A: 127.69, P: 45.2
7) A: 153.76, P: 49.6
8) A: 96.04, P: 39.2

Perimeter and Area of rectangles

1) A: 12.375, P: 15.5
2) A: 27.3, P: 21.4
3) A: 50.63, P: 28.8
4) A: 126.14, P: 45
5) A: 39.9, P: 25.7
6) A: 26.46, P: 21
7) A: 33.95, P: 26.4
8) A: 66.12, P: 34.4

Find the Area or Missing Side Length of a Rectangle

1) 33.66
2) 5.2
3) 7.2
4) 60.125
5) 100.125
6) 28.5
7) 10.2
8) 17.8

Area and Perimeter: Word Problems

1) 11.8
2) 10
3) 19.6
4) 10.24
5) 0.0576
6) 24.8

Volume of Cubes and Rectangle Prisms

1) 452.025 cm^3
2) 429.2 cm^3
3) 148.877 c m^3
4) 583.848 cm^3
5) 32
6) 40

Chapter 9 : Data Graphs, and Statistics

Topics that you'll learn in this chapter:

- ✓ Bar Graph,
- ✓ Tally and Pictographs,
- ✓ Dot Plots
- ✓ Line Graphs,
- ✓ Stem–And–Leaf Plot,
- ✓ Coordinate Plane,

Bar Graph

✍Graph the given information as a bar graph.

Day	Hot dogs sold
Monday	50
Tuesday	80
Wednesday	10
Thursday	30
Friday	70

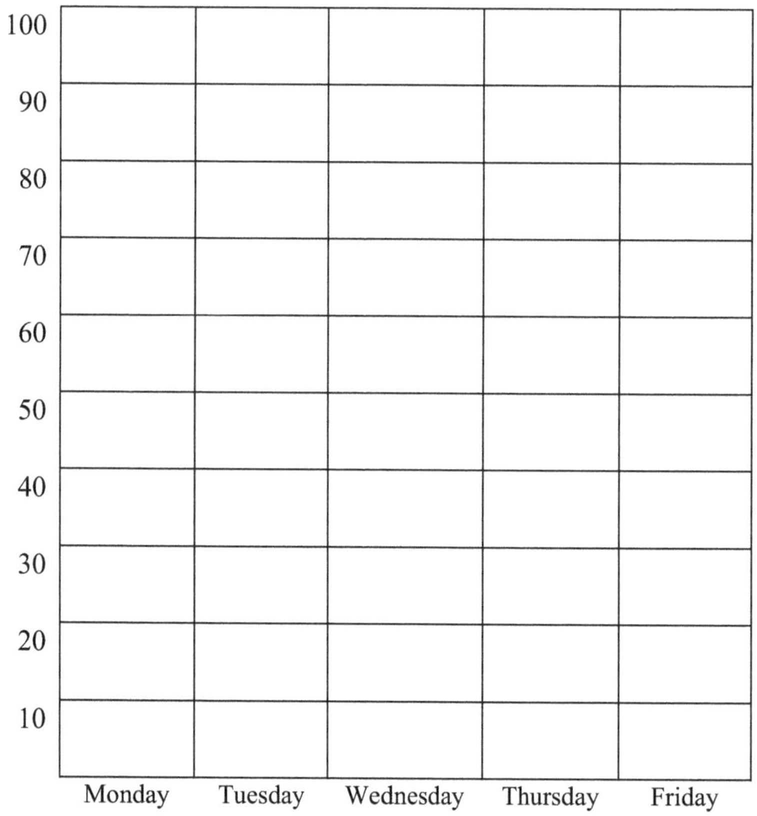

Tally and Pictographs

✍Using the key, draw the pictograph to show the information.

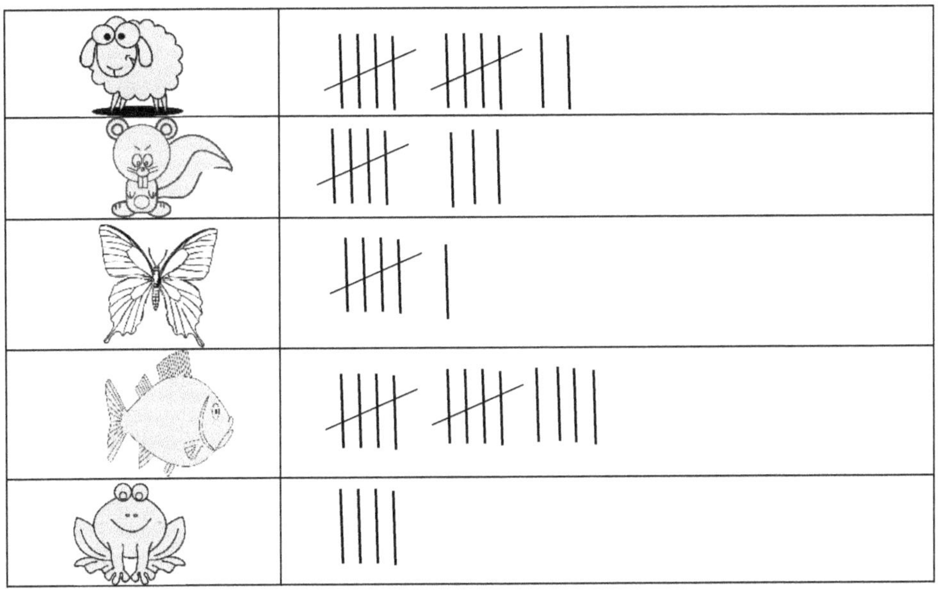

Key:  **= 2 animals**

Dot plots

The ages of students in a Math class are given below.

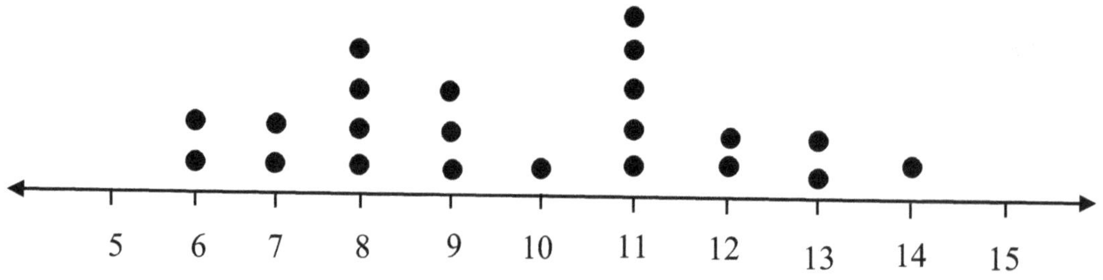

1) What is the total number of students in math class?
2) How many students are at least 12 years old?
3) Which age(s) has the most students?
4) Which age(s) has the fewest student?
5) Determine the median of the data.
6) Determine the range of the data.
7) Determine the mode of the data.

Line Graphs

David work as a salesman in a store. He records the number of shoes sold in five days on a line graph. Use the graph to answer the question.

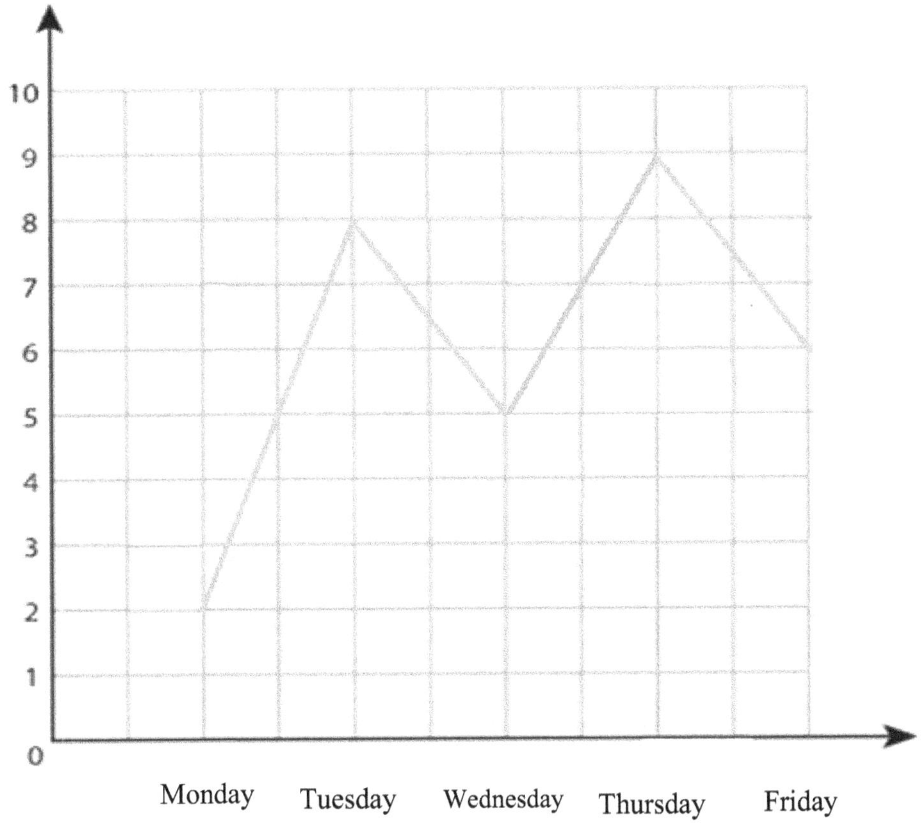

1) How many shoes were sold on Friday?
2) Which day had the minimum sales of shoes?
3) Which day had the maximum number of shoes sold?
4) How many shoes were sold in 5 days?

Stem–And–Leaf Plot

Make stem ad leaf plots for the given data.

1) 42, 47, 14, 19, 42, 69, 65, 49, 42, 10, 64

Stem | Leaf

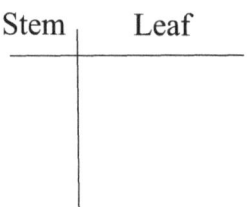

2) 43, 85, 52, 48, 45, 43, 51, 81, 59, 50, 85, 89

Stem | Leaf

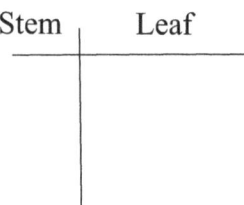

3) 112, 39, 46, 35, 80, 119, 42, 114, 37, 112, 47, 119

Stem | Leaf

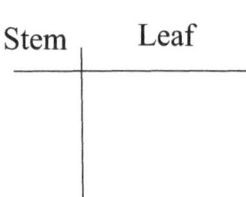

4) 90, 50, 131, 93, 112, 56, 139, 98, 115, 59, 98, 135, 111

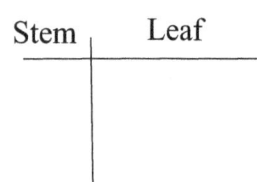

Stem | Leaf

Coordinate Plane

Plot each point on the coordinate grid.

1) A (4, 6)
2) B (3, 2)
3) C (1, 5)
4) D (5, 7)
5) E (4, 8)
6) F (9, 2)

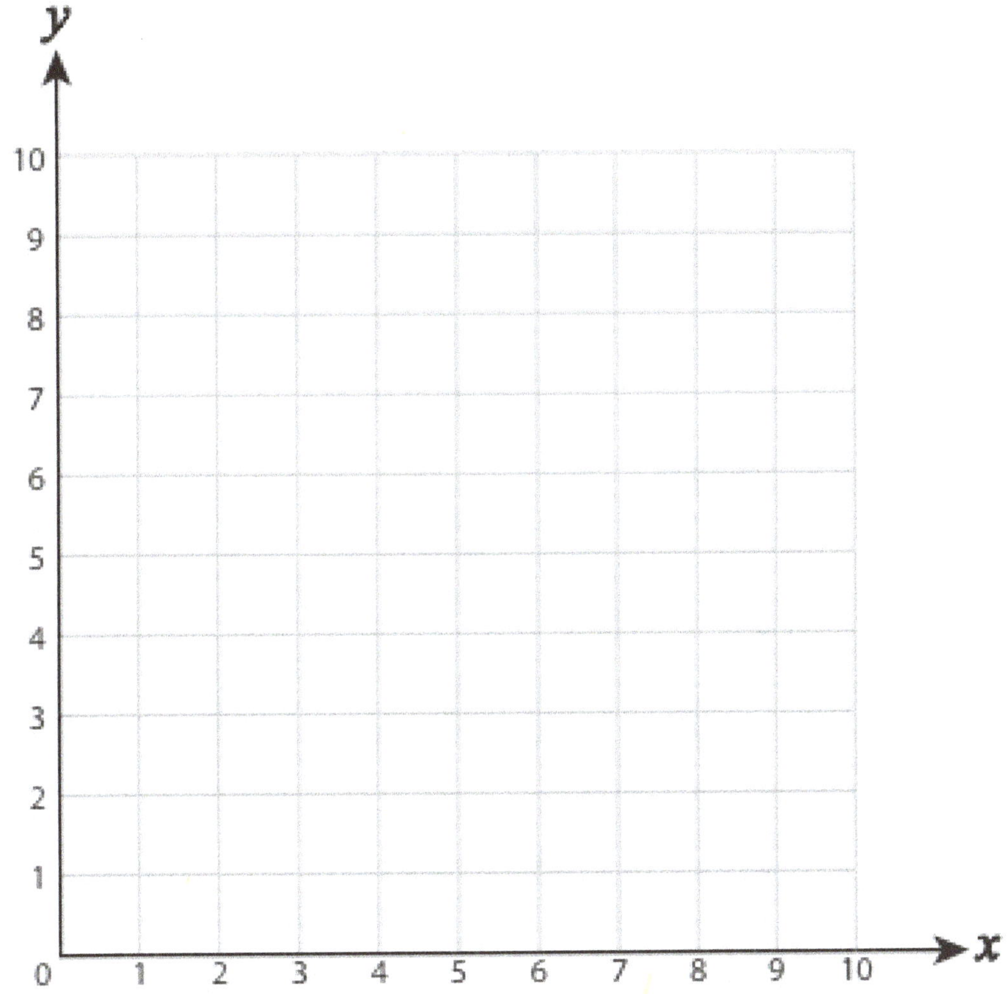

Answers of Worksheets

Bar Graph

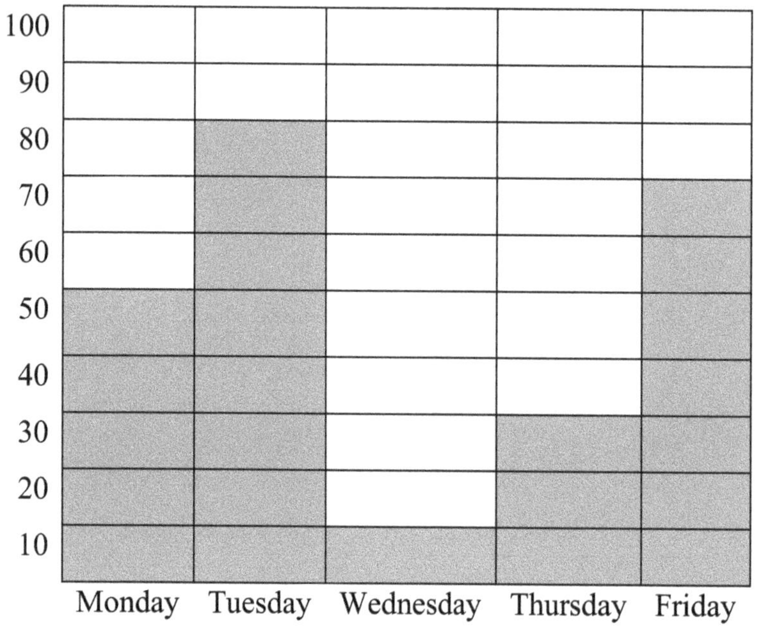

Tally and Pictographs

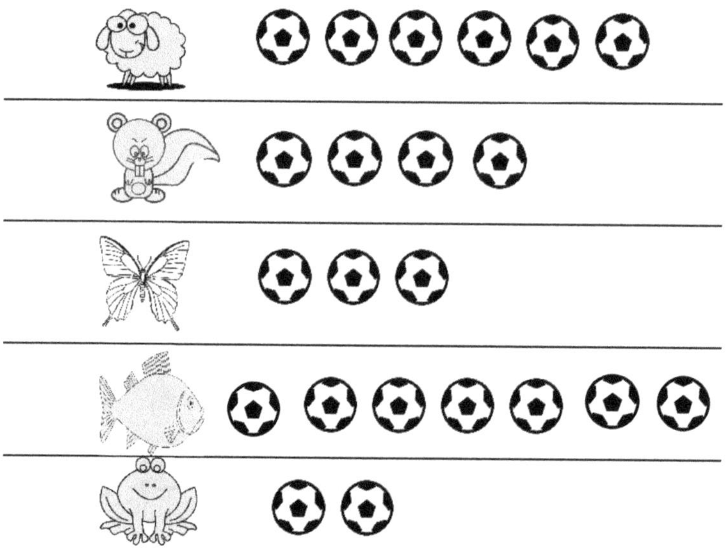

Dot plots

1) 22
2) 5
3) 11
4) 10 and 14
5) 2
6) 4
7) 2

Line Graphs

1) 6 2) Monday 3) Thursday 4) 30

Stem–And–Leaf Plot

1)

Stem	leaf
1	0 4 9
4	2 2 2 7 9
6	4 5 9

2)

Stem	leaf
4	3 3 5 8
5	0 1 2 9
8	1 5 5 9

3)

Stem	leaf
3	5 7 9
4	2 6 7
8	0
11	2 2 4 9 9

4)

Stem	leaf
5	0 6 9
9	0 3 8 8
11	1 2 5
13	1 5 9

Coordinate Plane

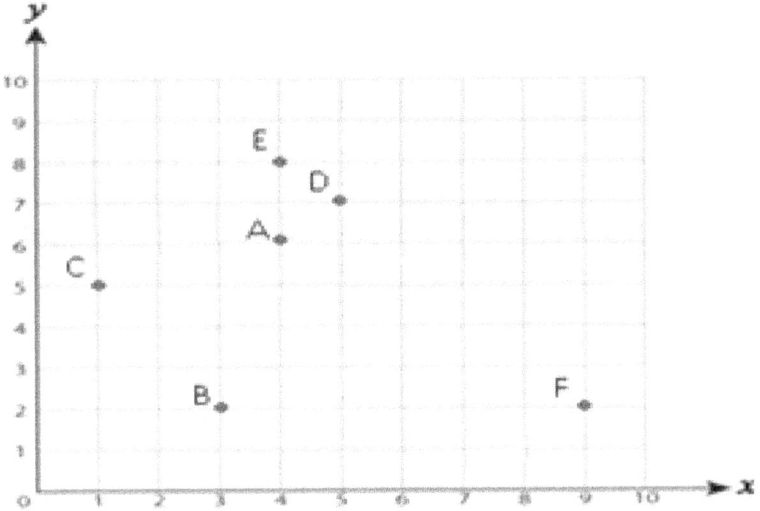

Chapter 10 : Three-Dimensional Figures

Topics that you'll learn in this chapter:

- ✓ Identify Three–Dimensional Figures,
- ✓ Count Vertices, Edges, and Faces,
- ✓ Identify Faces of Three–Dimensional Figures,

Identify Three–Dimensional Figures

Write the name of each shape.

1)

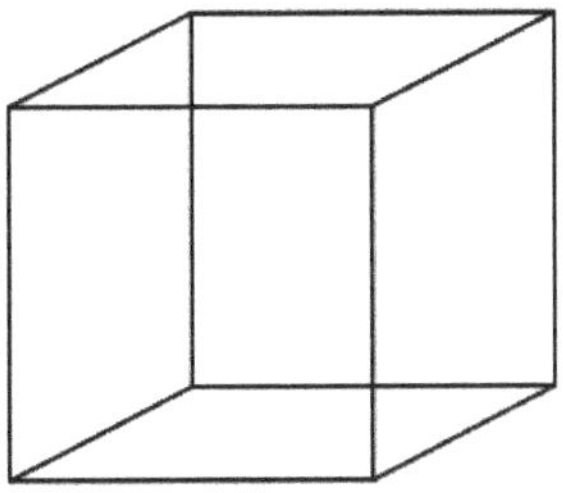

2)

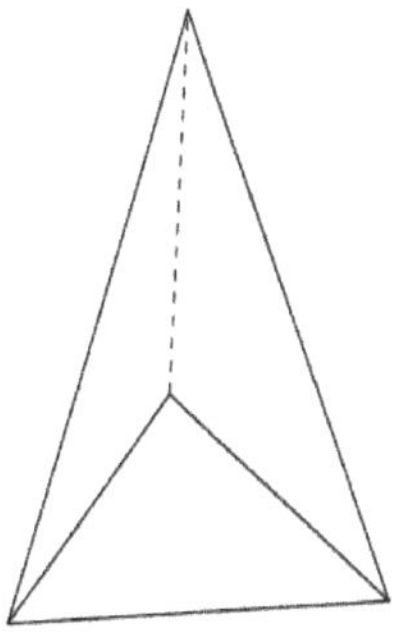

3)

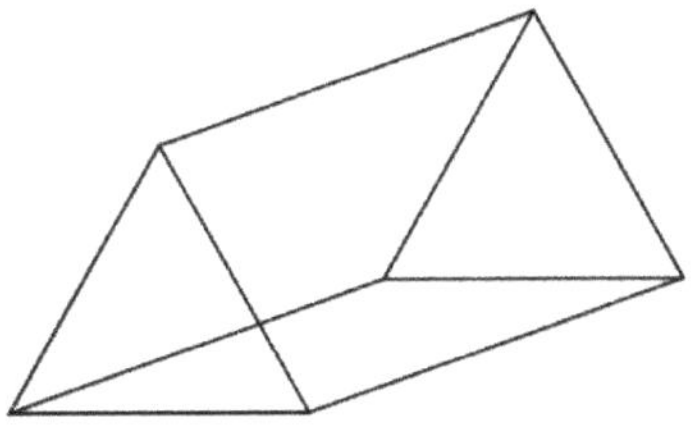

4)

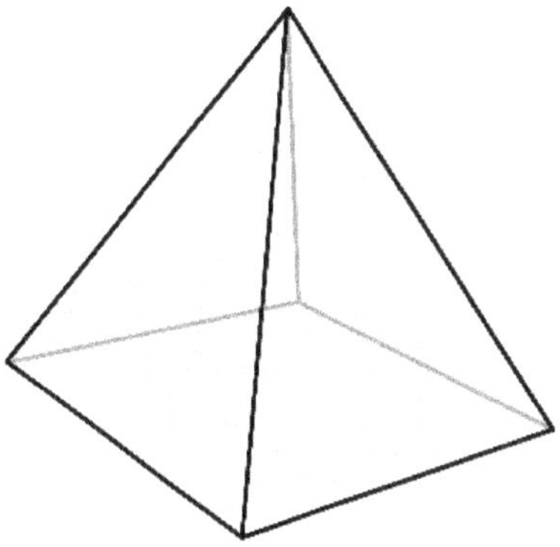

5)

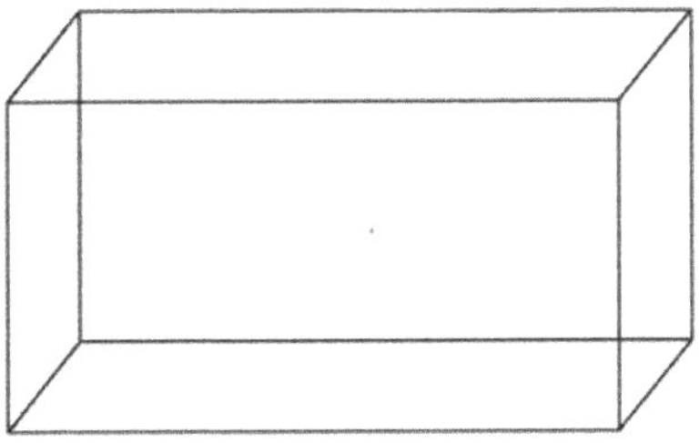

6)

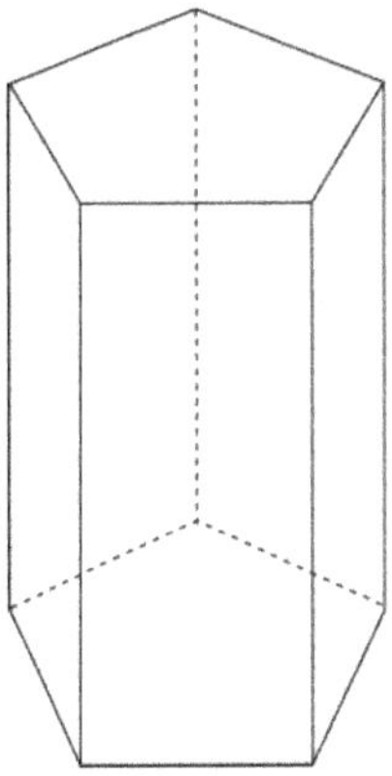

Count Vertices, Edges, and Faces

	Shape	Number of edges	Number of faces	Number of vertices
1)	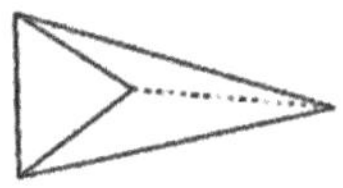	______	______	______
2)	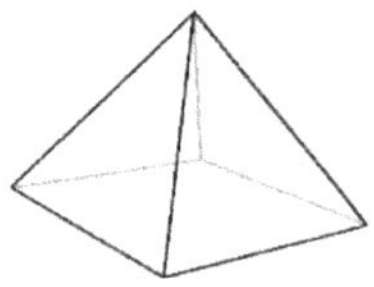	______	______	______
3)	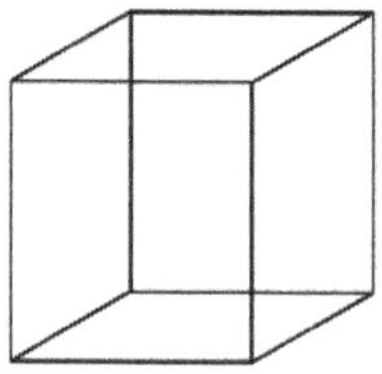	______	______	______
4)	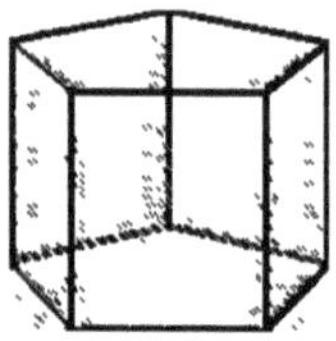	______	______	______
5)		______	______	______
6)	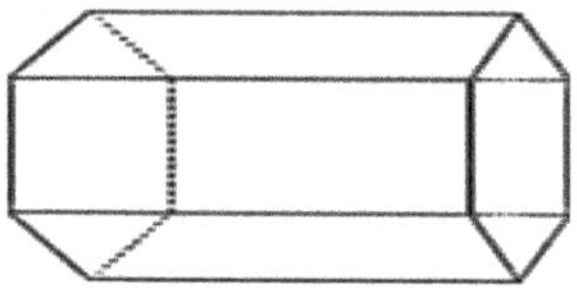	______	______	______

Identify Faces of Three–Dimensional Figures

Write the number of faces.

1)

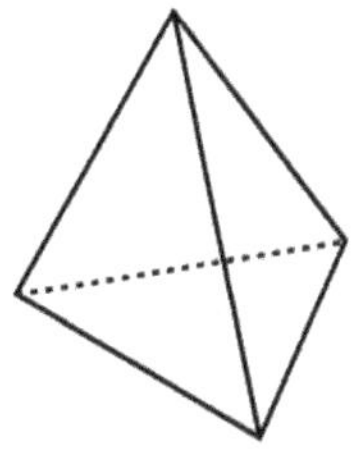

2)

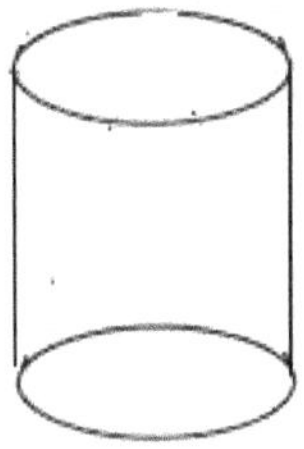

3)

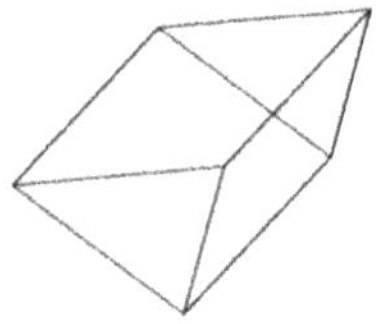

4)

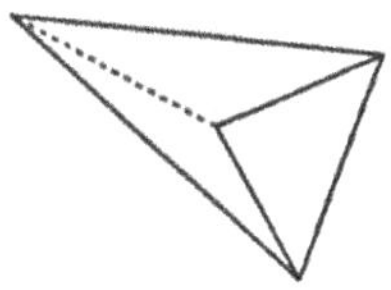

5)

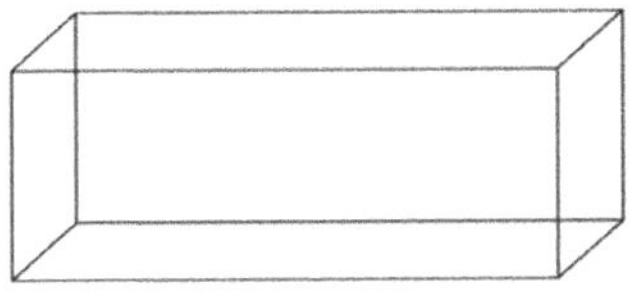

6)

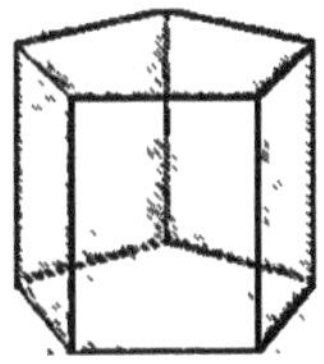

7)

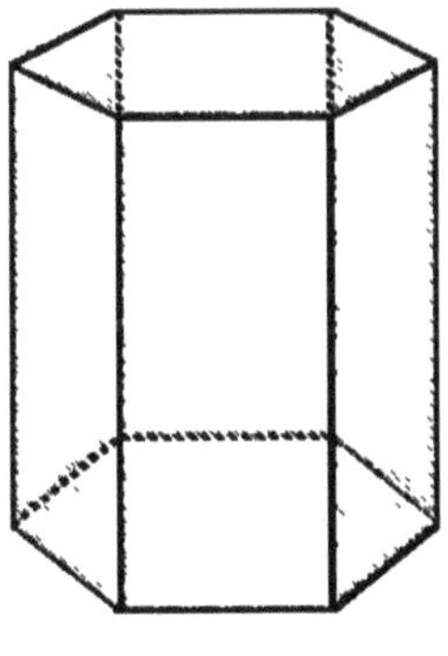

8)

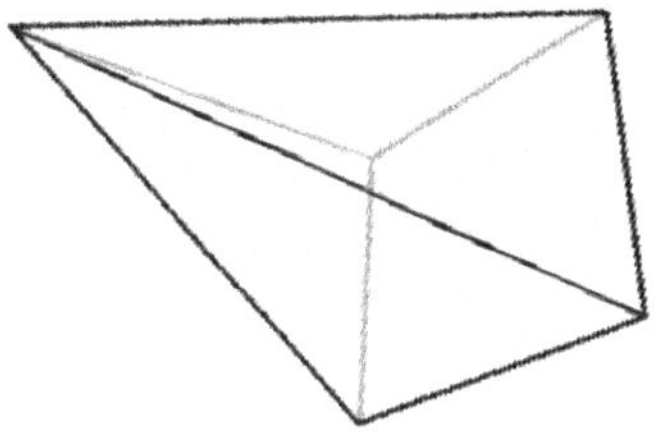

Answers of Worksheets

Identify Three–Dimensional Figures

1) Cube
2) Triangular pyramid
3) Triangular prism
4) Square pyramid
5) Rectangular prism
6) Pentagonal prism
7) Hexagonal prism

Count Vertices, Edges, and Faces

	Shape	Number of edges	Number of faces	Number of vertices
1)		6	4	4
2)	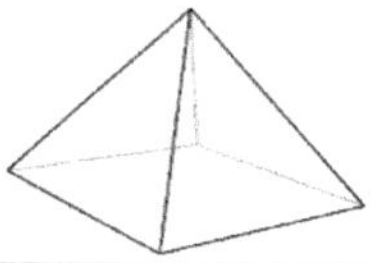	8	5	5
3)	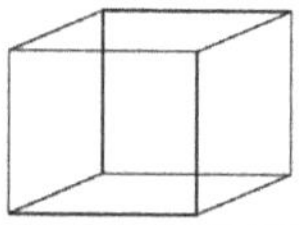	12	6	8
4)	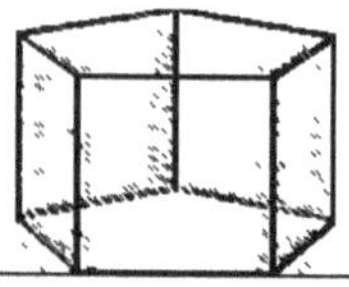	15	7	10
5)		12	6	8
6)	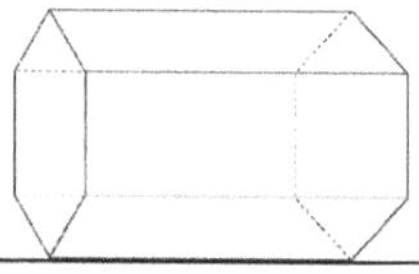	18	8	12

Identify Faces of Three–Dimensional Figures

1) 6
2) 2
3) 5
4) 4
5) 6
6) 7
7) 8
8) 5

Chapter 11 : Symmetry and Transformations

Topics that you'll learn in this chapter:

- ✓ Line Segments,
- ✓ Identify Lines of Symmetry,
- ✓ Count Lines of Symmetry,
- ✓ Parallel, Perpendicular and Intersecting Lines,

Line Segments

Write each as a line, ray, or line segment.

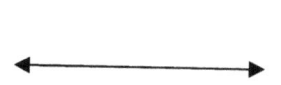

1)

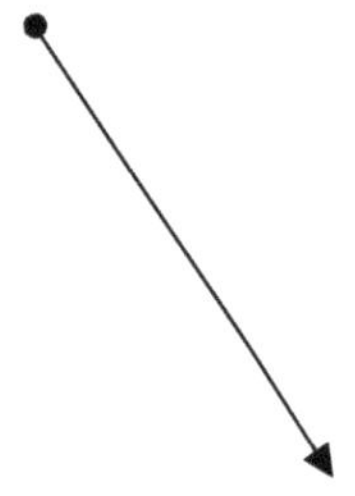

2)

3)

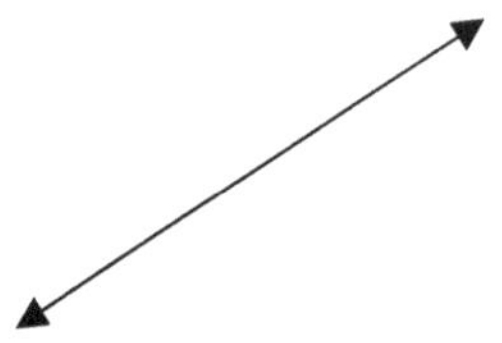

4)

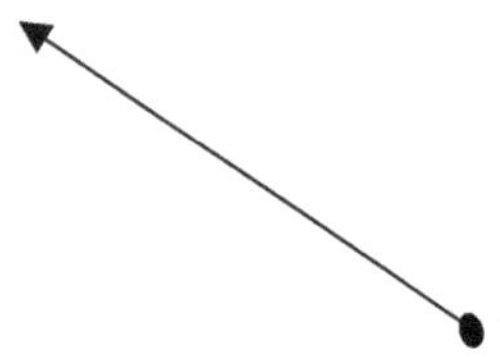

5)

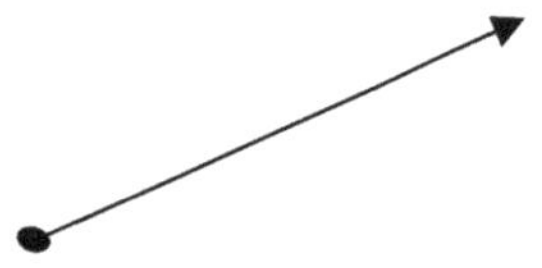

6)

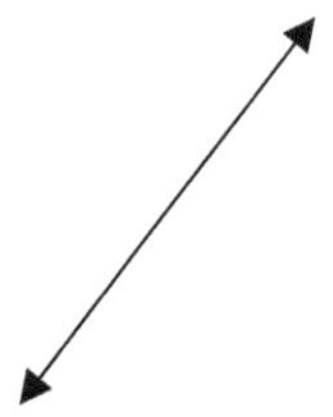

7)

8)

Identify Lines of Symmetry

Tell whether the line on each shape a line of symmetry is.

1)

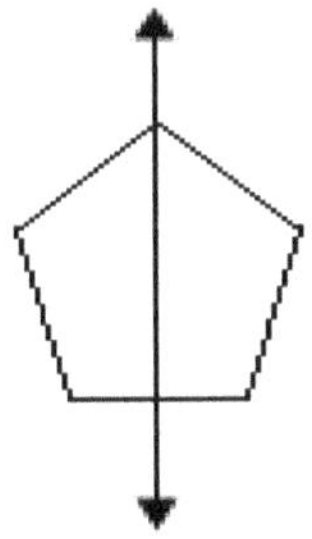

2)

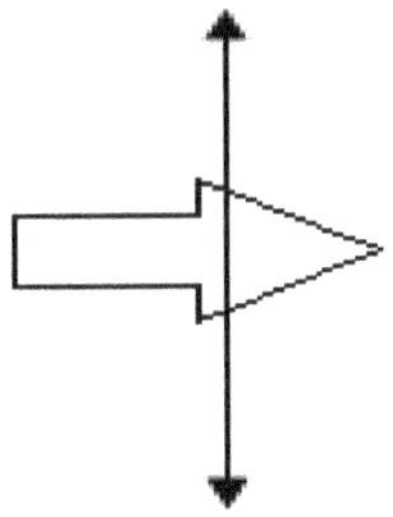

3)

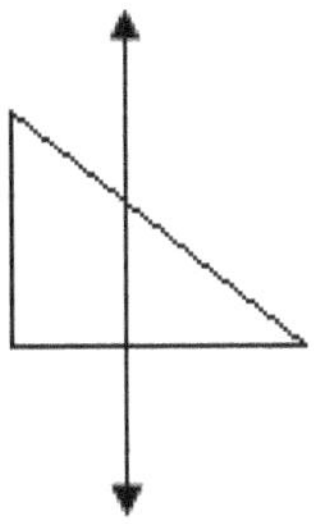

4)

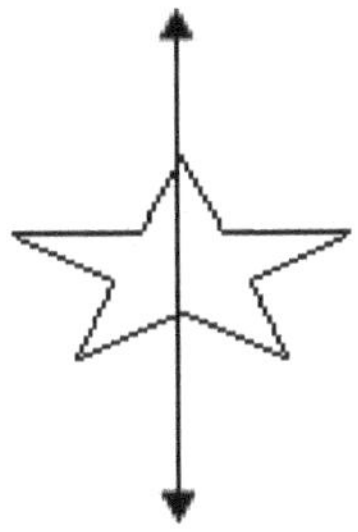

5)

6)

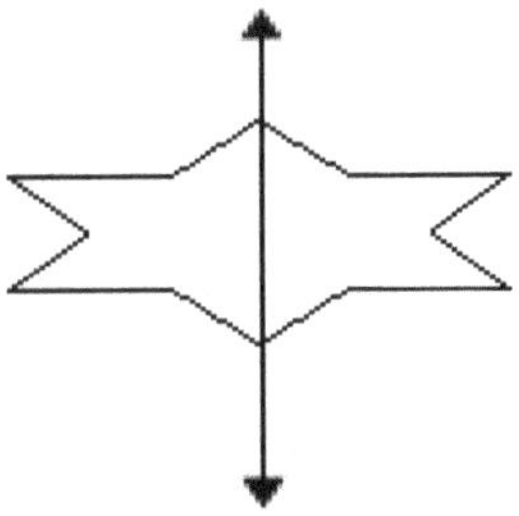

7)

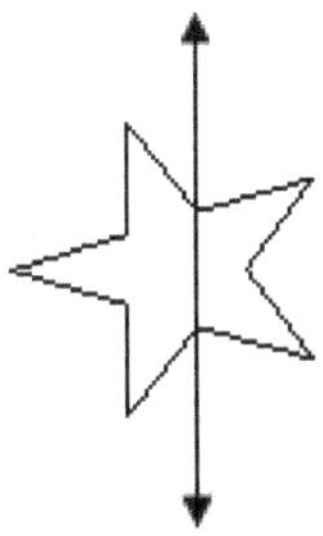

8)

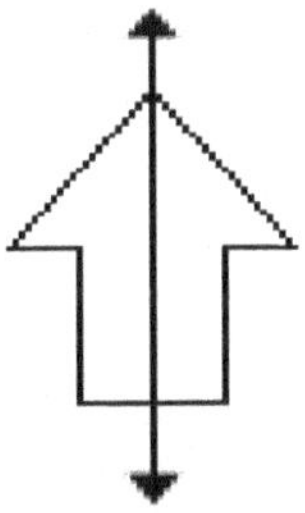

Count Lines of Symmetry

Draw lines of symmetry on each shape. Count and write the lines of symmetry you see.

1)

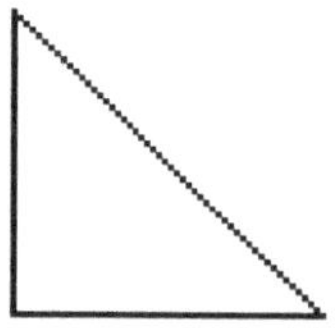

2)

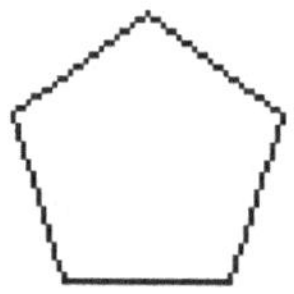

3)

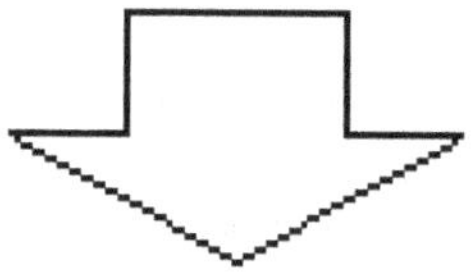

4)

5)

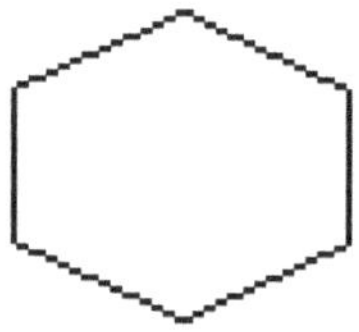

6)

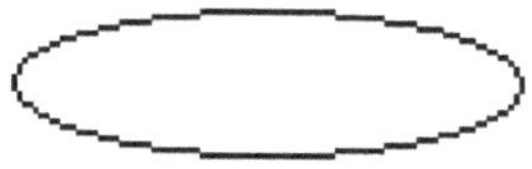

7)

8)

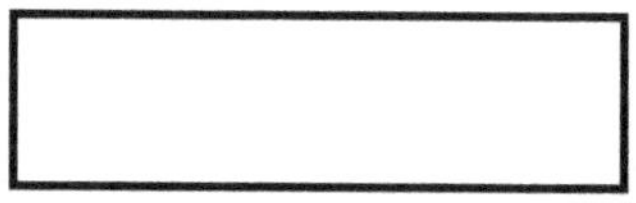

Parallel, Perpendicular and Intersecting Lines

State whether the given pair of lines are parallel, perpendicular, or intersecting.

1)

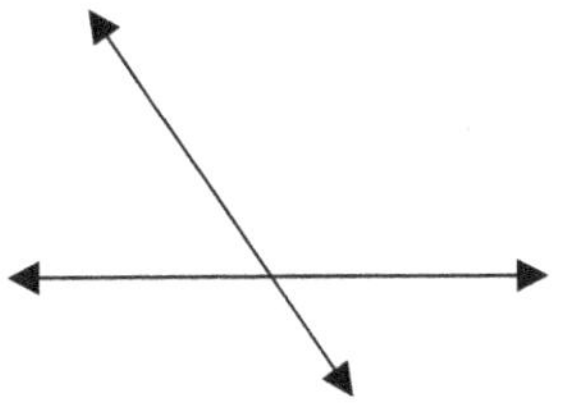

2)

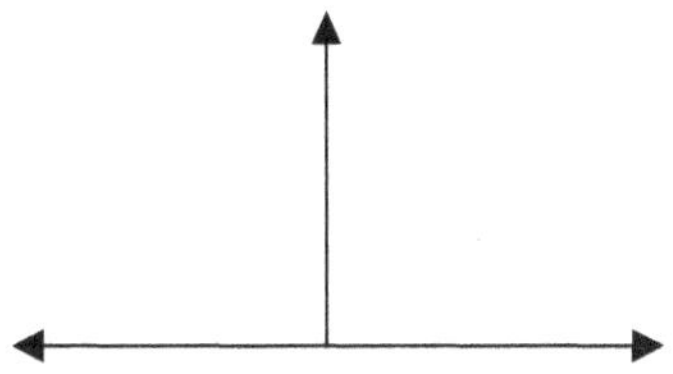

3)

4)

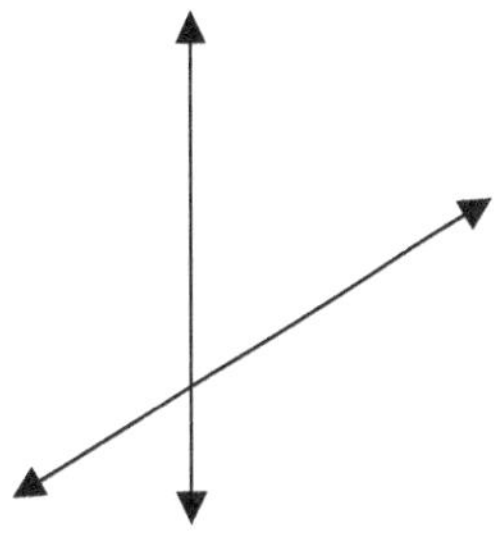

5)

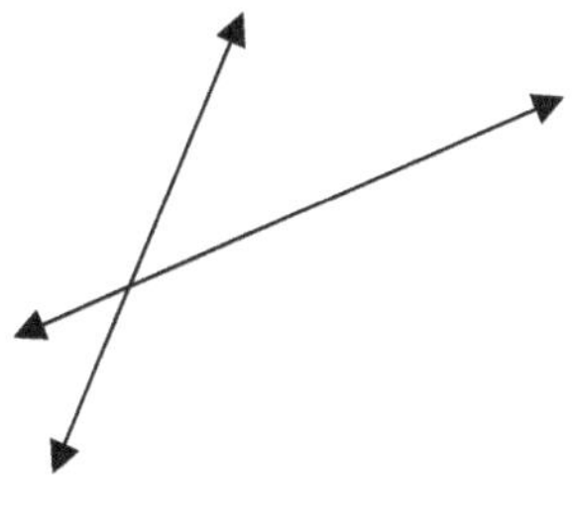

6)

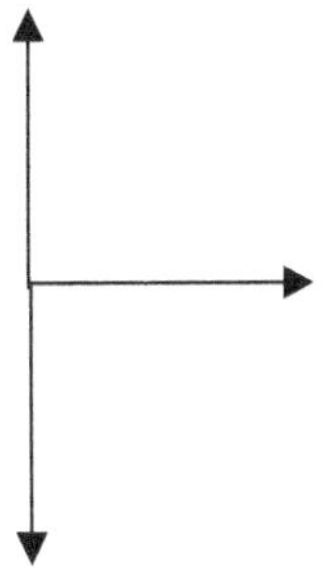

7)

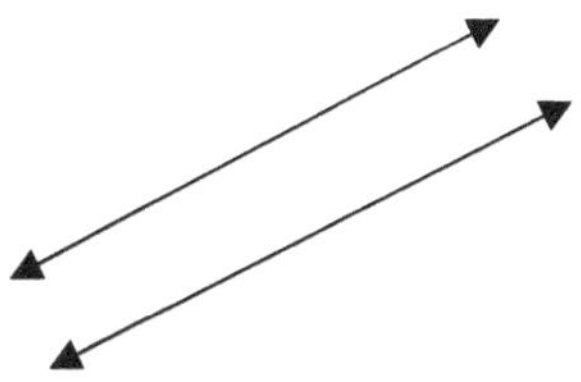

8)

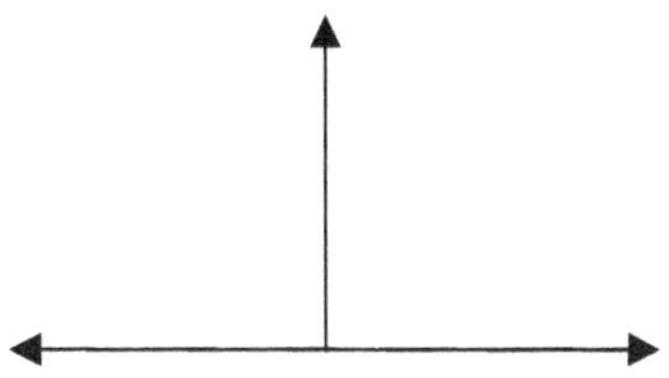

Answers of Worksheets

Line Segments

1) Ray
2) Line segment
3) Line
4) Ray
5) Ray
6) Line
7) Line
8) Line segment

Identify lines of symmetry.

1) yes
2) no
3) no
4) yes
5) yes
6) yes
7) no
8) yes

Count lines of symmetry.

1)

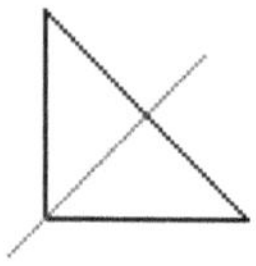

2)

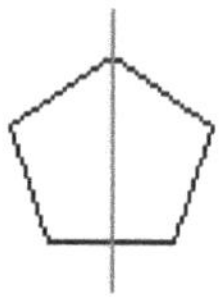

3)

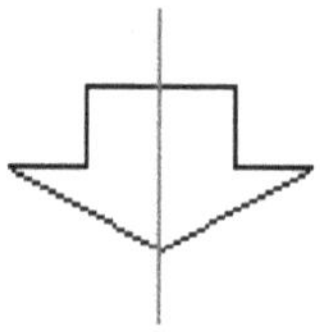

4)

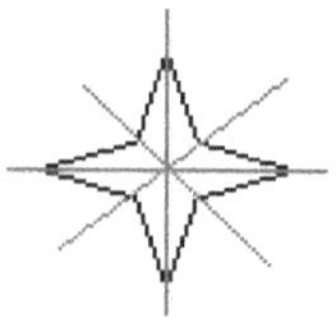

5)

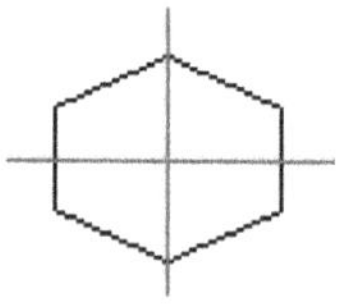

6)

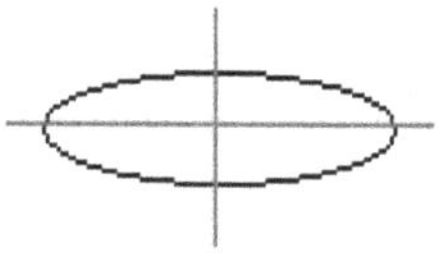

7)

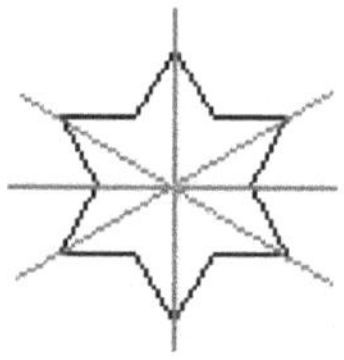

8)

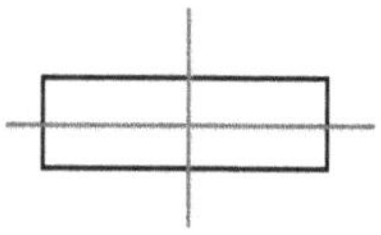

Parallel, Perpendicular and Intersecting Lines

1) Intersection
2) Perpendicular
3) Parallel
4) Intersection
5) Intersection
6) Perpendicular
7) Parallel
8) Perpendicular

Chapter 12 : Common Core Math Practice Tests

Time to Test

Time to refine your skill with a practice examination.

Take a REAL Common Core Mathematics test to simulate the test day experience. After you've finished, score your test using the answer key.

Before You Start

- You'll need a pencil and scratch papers to take the test.
- For this practice test, don't time yourself. Spend time as much as you need.
- It's okay to guess. You won't lose any points if you're wrong.
- After you've finished the test, review the answer key to see where you went wrong.

Calculators are not permitted for Grade 4 Common Core Tests

Good Luck!

Common Core GRADE 4 MAHEMATICS REFRENCE MATERIALS

LENGTH

Customary	Metric
1 mile (mi) = 1,760 yards (yd)	1 kilometer (km) = 1,000 meters (m)
1 yard (yd) = 3 feet (ft)	1 meter (m) = 100 centimeters (cm)
1 foot (ft) = 12 inches (in.)	1 centimeter (cm) = 10 millimeters (mm)

VOLUME AND CAPACITY

Customary	Metric
1 gallon (gal) = 4 quarts (qt)	1 liter (L) = 1,000 milliliters (mL)
1 quart (qt) = 2 pints (pt.)	
1 pint (pt.) = 2 cups (c)	
1 cup (c) = 8 fluid ounces (Fl oz)	

WEIGHT AND MASS

Customary	Metric
1 ton (T) = 2,000 pounds (lb.)	1 kilogram (kg) = 1,000 grams (g)
1 pound (lb.) = 16 ounces (oz)	1 gram (g) = 1,000 milligrams (mg)

Time

1 year = 12 months	1 day = 24 hours
1 year = 52 weeks	1 hour = 60 minutes
1 week = 7 days	1 minute = 60 seconds

Perimeter

Square			$P = 4S$
Rectangle	$P = L + W + L + W$	or	$P = 2L + 2W$

Area

Square	$A = S \times S$
Rectangle	$A = L \times W$

Common Core Practice Test 1

Mathematics

GRADE 4

Released Month Year

1) Which number correctly completes the number sentence $50 \times 32 = ?$

A. 3,600

B. 800

C. 1,800

D. 1,600

2) Tam has 350 cards. He wants to put them in boxes of 70 cards. How many boxes does he need?

A. 3

B. 5

C. 8

D. 6

3) If this clock shows a time in the noon, what time was it 2 hours and 45 minutes ago?

A. 3:45 AM

B. 3:30 AM

C. 2:30 AM

D. 2:45 AM

4) Joe has 679 crayons. What is this number rounded to the nearest ten?

(Write your answer in the box below).

5) Use a rule to measure the length and width of the following rectangle to the nearest inches.

What measurement is the closest to the area of the rectangle in square inches?

A. 12 square inches

B. 15 square inches

C. 28 square inches

D. 10 square inches

6) There are 7 days in a week. There are 31 days in the month of July. How many times as many days are there in July than are in one week?

A. 5 times

B. 6 times

C. 4 times

D. 3 times

7) A football team is buying new uniforms. Each uniform cost \$19. The team wants to buy 15 uniforms.

Which equation represents a way to find the total cost of the uniforms?

A. $(19 \times 15) + (10 \times 5) = 285 + 50$

B. $(19 \times 6) + (10 \times 15) = 109 + 150$

C. $(19 \times 10) + (19 \times 5) = 190 + 95$

D. $(19 \times 10) + (5 \times 15) = 190 + 75$

8) A number sentence such as $20 + Z = 55$ can be called an equation. If this equation is true, then which of the following equations is not true?

A. $55 - 20 = Z$

B. $55 - Z = 20$

C. $Z - 20 = 55$

D. $Z + 20 = 55$

9) Ella described a number using these clues:

Three – digit even numbers that have a 5 in the hundreds place and a 3 in the tens place. Which number could fit Ella's description?

A. 533

B. 532

C. 535

D. 537

10) Use the table below to answer the question.

Favorite Sports

Sport	Number of Votes
football (FB)	9
basketball (BB)	4
soccer (SOC)	10
volleyball (VB)	2

The students in the fourth-grade class voted for their favorite sport. Which bar graph shows results of the students vote?

A.

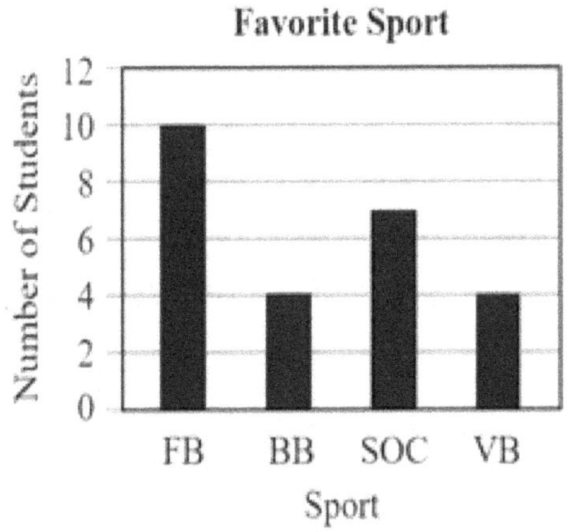

B.

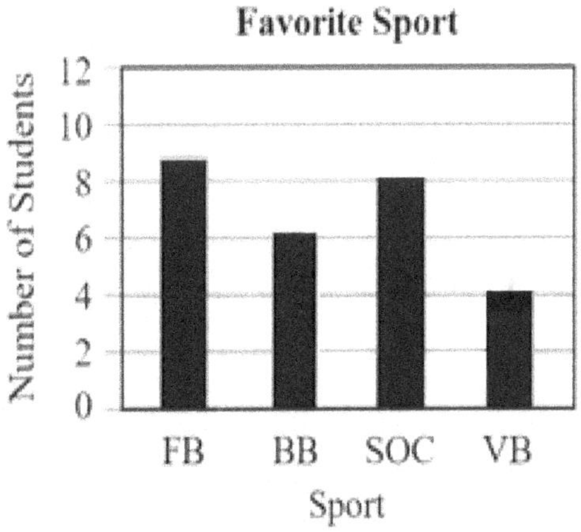

C.

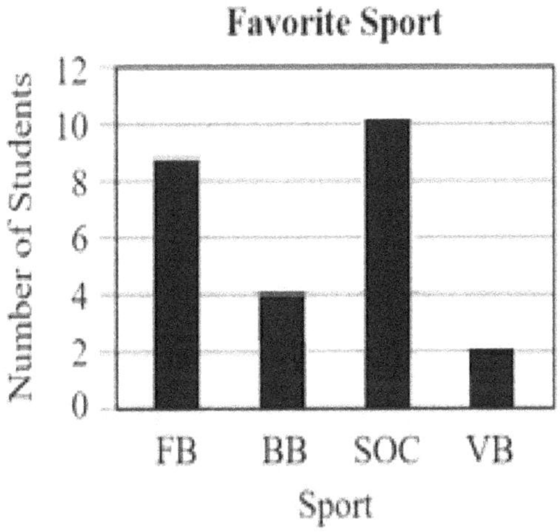

D.

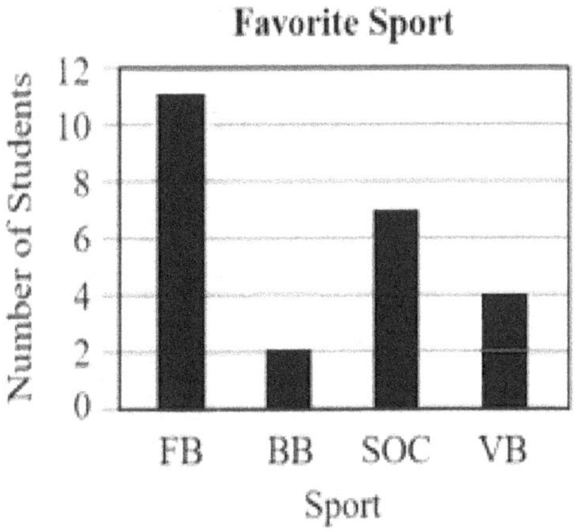

11) Use the picture below to answer the question.

Which decimal number names the shaded part of this square?

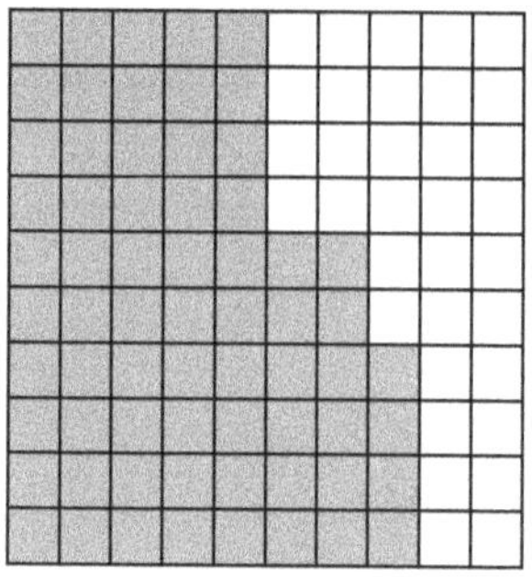

A. 0.32

B. 0.68

C. 0.34

D. 0.66

12) Use the table below to answer the question.

Which list of city populations is in order from least to greatest?

A. 30,032; 21,851; 37,009; 28,081.

B. 21,851; 28,081; 30,032; 37,009.

C. 28,081; 30,032; 37,009; 21,851.

D. 37,009; 30,032; 28,081; 21,851.

City Populations	
City	**Population**
Denton	21,851
Bomberg	30,032
Windham	28,081
Sanhill	37,009

13) Which number correctly completes the subtraction sentence 6.0 – 4.75 = _____?

A. 1.15

B. 1.25

C. 1.50

D. 2.50

14) For a concert, there are children's tickets and adult tickets for sale. Of the total available tickets, $\frac{52}{100}$ have been sold as adult tickets and $\frac{1}{10}$ as children's tickets. The rest of the tickets have not been sold.

What fraction of the total number of tickets for the concert have been sold?

A. $1\frac{1}{52}$

B. $\frac{31}{50}$

C. $\frac{31}{52}$

D. $\frac{52}{10}$

15) Circle a reasonable measurement for the angle F:

A. 125°

B. 60°

C. 45°

D. 190°

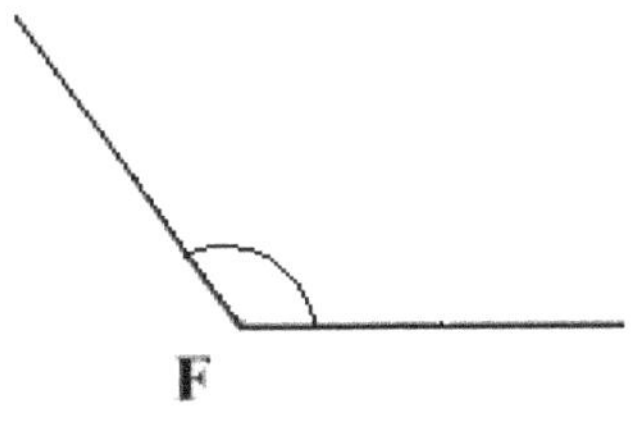

16) Peter's pencil is $\frac{51}{10}$ of a meter long. What is the length, in meters, of Peter's pencil written as a decimal?

A. 5.1

B. 0.051

C. 0.51

D. 51.10

17) Mia has a group of shapes. Each shape in her group has at least one set of parallel sides. Each shape also has at least one set of perpendicular sides. Which group could be Mia's group of shapes?

A.

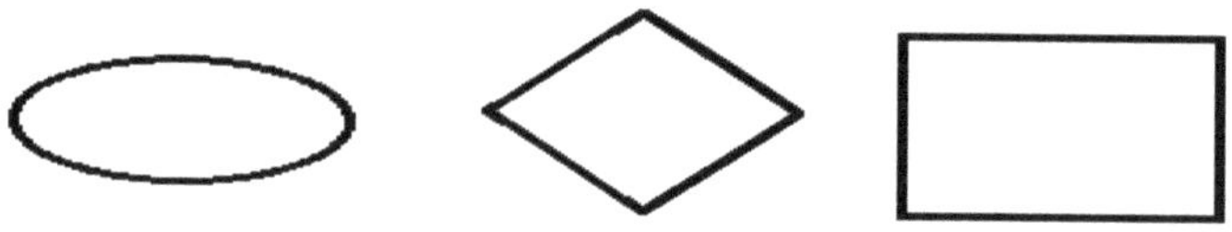

B.

C.

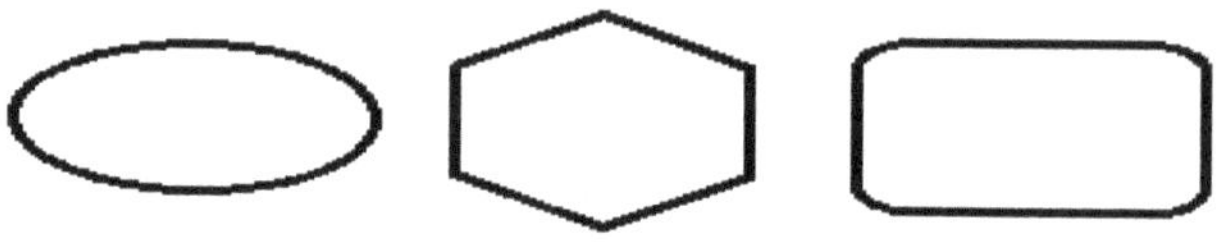

D.

18) There are 58 students from Riddle Elementary school at the library on Monday. The other 17 students in the school are practicing in the classroom. Which number sentence shows the total number of students in Riddle Elementary school?

A. $58 + 17$

B. $58 - 17$

C. 58×17

D. $58 \div 17$

19) A stack of 4 pennies has a height of 1 centimeter. Elise has a stack of pennies with a height of 16 centimeters. Which equation can be used to find the number of pennies, n, in Elise's stack of pennies?

A. $n = 16 + 4$

B. $n = 16 - 4$

C. $n = 4 \times 16$

D. $n = 16 \div 34$

20) Use the models below to answer the question.

Which statement about the models is true?

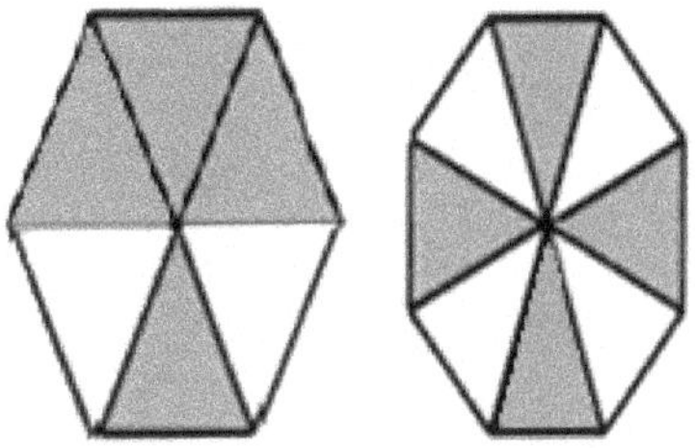

A. Each shows the same fraction because they are the same size.

B. Each shows a different fraction because they are different shapes.

C. Each shows the same fraction because they both have 4 sections shaded.

D. Each shows a different fraction because they both have 4 shaded sections but a different number of total sections.

21) What is the value of A in the equation $24 \div A = 4$?

A. 2

B. 6

C. 7

D. 8

22) Sophia flew 7,143 miles from Los Angeles to New York City. What is the number of miles Sophia flew rounded to the nearest thousand?

A. 7,000

B. 7,100

C. 7,140

D. 8,000

23) Write $\frac{75}{100}$ as a decimal number.

A. 7.05

B. 0.75

C. 7.5

D. 0.075

24) Erik made 12 pints of juice. He drinks 2 cups of juice each day. How many days will Erik take to drink all of the juice he made?

A. 14 days

B. 6 days

C. 10 days

D. 4 days

25) Jason has prepared $\frac{8}{1,000}$ of his assignment. Which decimal represent the part of the assignment Jason has prepared?

A. 8.110

B. 8.001

C. 0.008

D. 0.0008

26) Emma described a number using these clues.

✓ 3 digits of the number are 9, 3, and 2

✓ The value of the digit 3 is (3×10)

✓ The value of the digit 2 is $(2 \times 1,000)$

✓ The value of the digit 9 is $(9 \times 10,000)$

Which number could fit Emma's description?

A. 92,563.10

B. 29,563.70

C. 92,536.17

D. 92,635.71

27) There are 12 boxes and each box contain 24 pencils. How many pencils are in the boxed in total?

A. 188

B. 208

C. 288

D. Not here

28) Emily and Ava were working on a group project last week. They completed $\frac{7}{17}$ of their project on Tuesday and the rest on Wednesday.

Ava completed $\frac{6}{17}$ of their project on Tuesday. What fraction of the group project did Emily completed on Tuesday?

A. $\frac{2}{17}$

B. $\frac{13}{17}$

C. $\frac{1}{17}$

D. $\frac{5}{17}$

29) Moe packs 50 boxes with flashcards. Each box holds 80 flashcards. How many flashcards Moe can pack into these boxes?

A. 4,000

B. 400

C. 700

D. 4,400

30) Jason's favorite sports team has won 0.13 of its games this season. How can Jason express this decimal as a fraction?

A. $\frac{1}{13}$

B. $\frac{13}{10}$

C. $\frac{13}{100}$

D. $\frac{0.13}{100}$

"This is the end of Practice Test 1"

Common Core Practice Test 2

Mathematics

GRADE 4

Released Month Year

1) Which statement about the number 495,271.38 is true?

A. The digit 9 has a value of $(9 \times 1{,}000)$

B. The digit 5 has a value of (5×100)

C. The digit 7 has a value of (7×10)

D. The digit 4 has a value of $(4 \times 10{,}000)$

2) What is the perimeter of this rectangle?

A. 16 cm

B. 28 cm

C. 14 cm

D. 24 cm

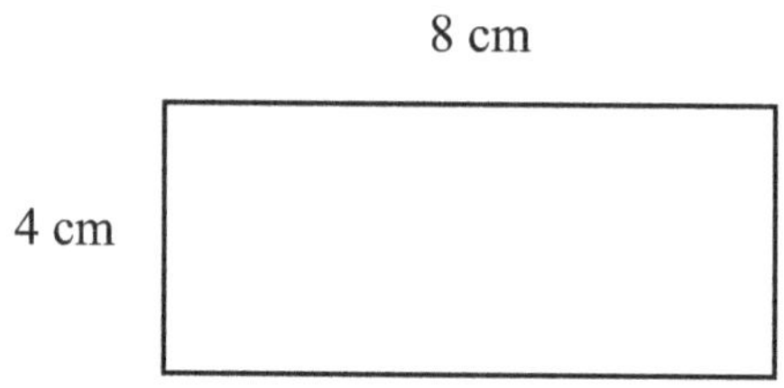

3) What is the eighth number in the following pattern?

1,450; 1,700; 1,950; 2,200; _____, _____, _____, _____

A. 3,400

B. 3,150

C. 3,350

D. 3,200

4) Jamie has 4 quarters, 8 dime, and 25 pennies. How much money does Jamie have?

A. 195 pennies

B. 205 pennies

C. 205 pennies

D. 190 pennies

5) Jeb paid $96 for a magazine subscription. If he is paying $8 for each issue of the magazine, how many issues of the magazine will he receive?

A. 12

B. 14

C. 24

D. 10

6) What is the perimeter of the triangle?

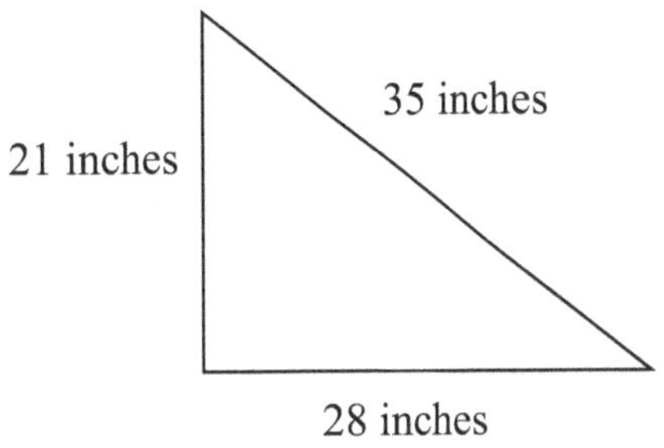

A. 46 inches

B. 68 inches

C. 84 inches

D. 92 inches

7) Emma draws a shape on her paper. The shape has four sides. It has both diagonals have equal measure and are perpendicular. What shape does Emma draw?

A. parallelogram

B. rectangle

C. square

D. trapezoid

8) Order the fractions from least to greatest. $\frac{5}{8}, \frac{1}{2}, \frac{5}{6}, \frac{1}{24}$

A. $\frac{5}{8}, \frac{5}{6}, \frac{1}{2}, \frac{1}{24}$

B. $\frac{5}{6}, \frac{5}{8}, \frac{1}{24}, \frac{1}{2}$

C. $\frac{1}{24}, \frac{1}{2}, \frac{5}{8}, \frac{5}{6}$

D. $\frac{1}{2}, \frac{1}{24}, \frac{5}{6}, \frac{5}{8}$

9) There were 52 students in the first row and 9 students in the second row. How many students were in the first two rows?

A. 61

B. 57

C. 43

D. 55

10) Jack has \$31.00. He earns \$14.00 more. How much money does Jack have in all?

A. \$54.00

B. \$44.00

C. \$46.00

D. \$45.00

11) Joe put 15 red cards and 6 black cards in each bag. What is the total number of cards Joe put in 4 bags?

Write your answer in the box below.

12) Rounded to the nearest 100,000, the population of Louisiana was 54,700,000 in 2010. Which number could be the actual population of Louisiana in 2010?

A. 54,789,221

B. 54,752,220

C. 54,723,330

D. 54,762,990

13) A football teams play 48 games each year. How many games will the team play in 5 years?

A. 260

B. 180

C. 240

D. 1,100

14) The figure below shows a diagram of a living room.

The perimeter of the living room is 26 feet (ft). What is the width(w) of the living room?

A. 12 ft

B. 7 ft

C. 5 ft

D. 11 ft

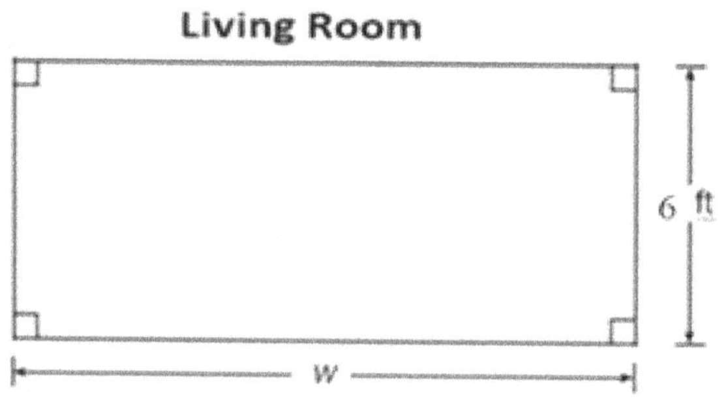

15) A building is 63 feet high. What is the height of the building in yards?

A. 22 yards

B. 21 yards

C. 23 yards

D. 66 yards

16) The sum of A and B equals 35. If A = 18, which equation can be used to find the value of B?

A. B – 18 = 35

B. B + 18 = 35

C. A + 18 = 35

D. A – 18 = 35

17) Which number is represented by A?$4 \times A = 172$

A. 43

B. 37

C. 47

D. 31

18) Which fraction has the least value?

A. $\frac{1}{32}$

B. $\frac{1}{2}$

C. $\frac{3}{4}$

D. $\frac{5}{8}$

19) Which triangle has one obtuse angle?

A.

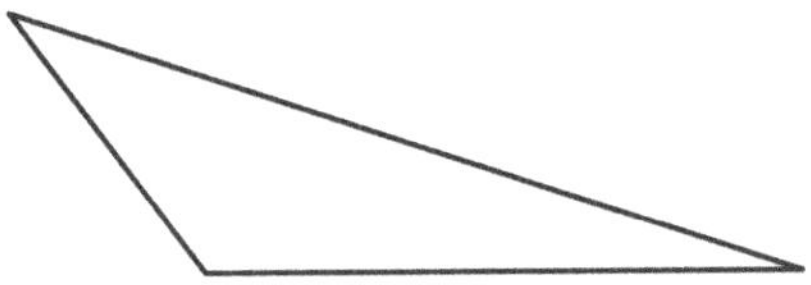

C.

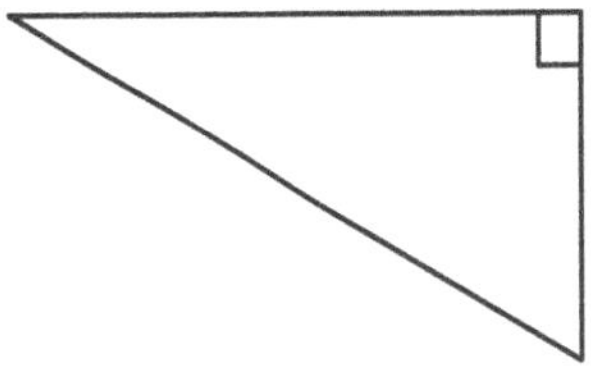

B.

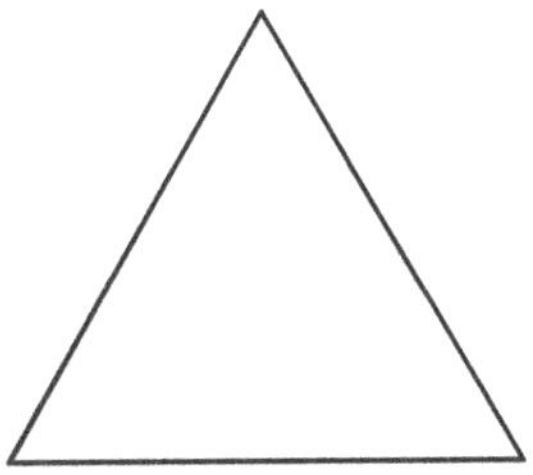

D.

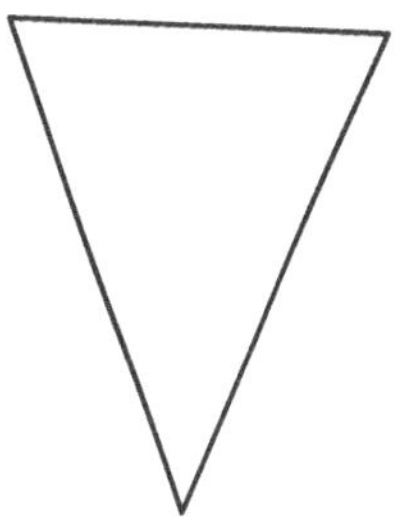

20) Lisa has 64 pastilles. She wants to put them in boxes of 16 pastilles. How many boxes does she need?

A. 8

B. 5

C. 6

D. 4

21) What is the volume of the cube?

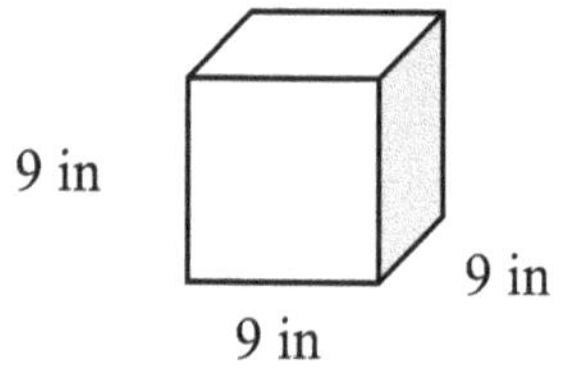

Write your answer in the box below.

22) To what number is the arrow pointing?

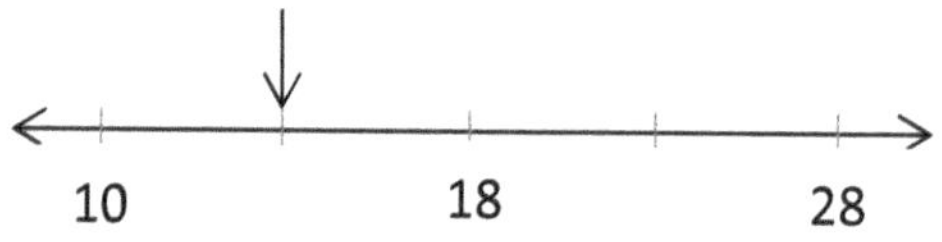

A. 16

B. 14

C. 12

D. 22

23) What mixed number is shown by the shaded rectangles?

A. $3\frac{3}{4}$

B. $3\frac{1}{4}$

C. $2\frac{1}{4}$

D. $2\frac{3}{4}$

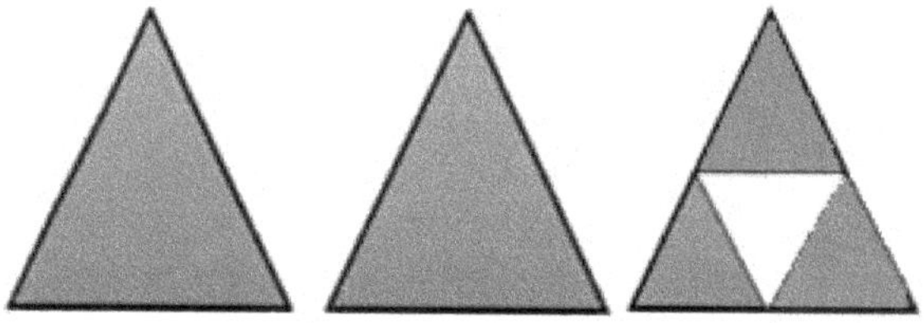

17) Which number is represented by A?$4 \times A = 172$

A. 43

B. 37

C. 47

D. 31

18) Which fraction has the least value?

A. $\frac{1}{32}$

B. $\frac{1}{2}$

C. $\frac{3}{4}$

D. $\frac{5}{8}$

19) Which triangle has one obtuse angle?

A.

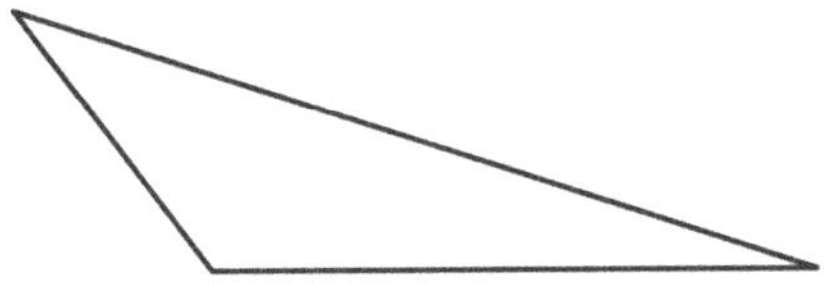

C.

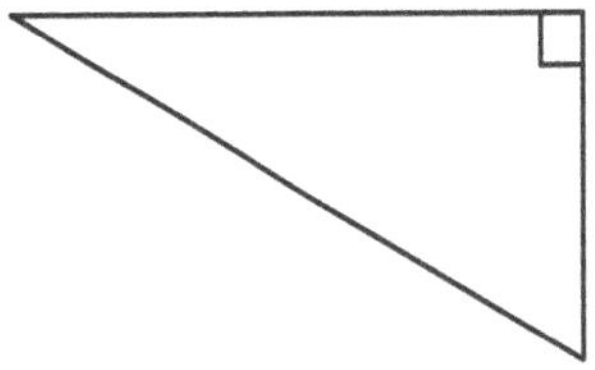

B.

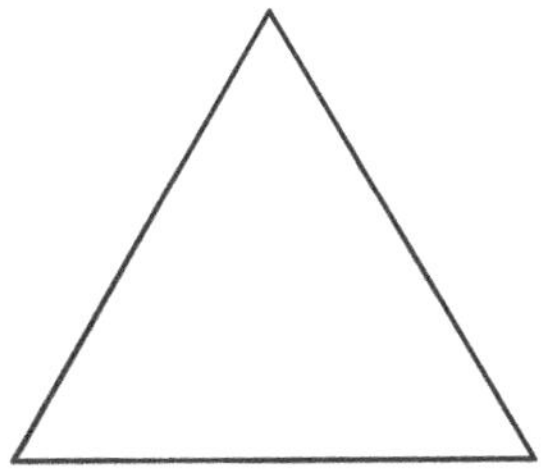

D.

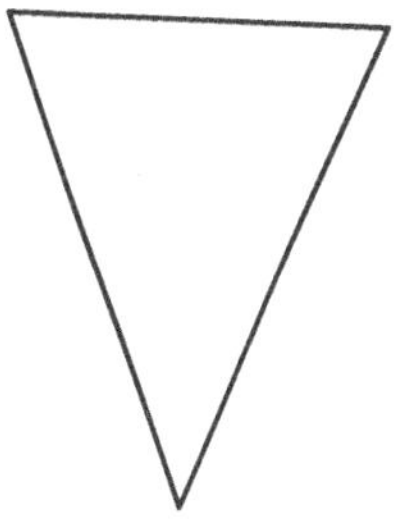

20) Lisa has 64 pastilles. She wants to put them in boxes of 16 pastilles. How many boxes does she need?

A. 8

B. 5

C. 6

D. 4

21) What is the volume of the cube?

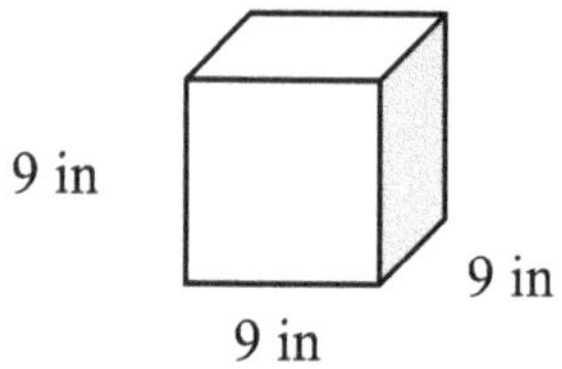

Write your answer in the box below.

22) To what number is the arrow pointing?

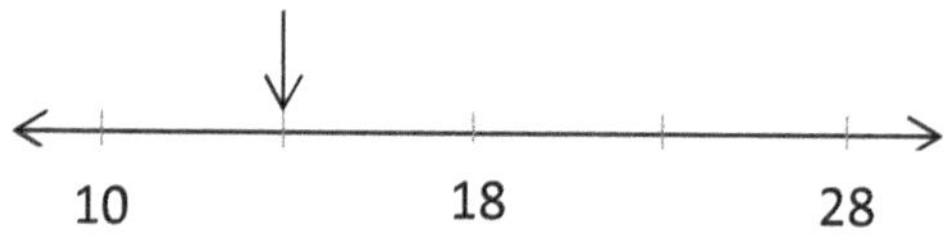

A. 16

B. 14

C. 12

D. 22

23) What mixed number is shown by the shaded rectangles?

A. $3\frac{3}{4}$

B. $3\frac{1}{4}$

C. $2\frac{1}{4}$

D. $2\frac{3}{4}$

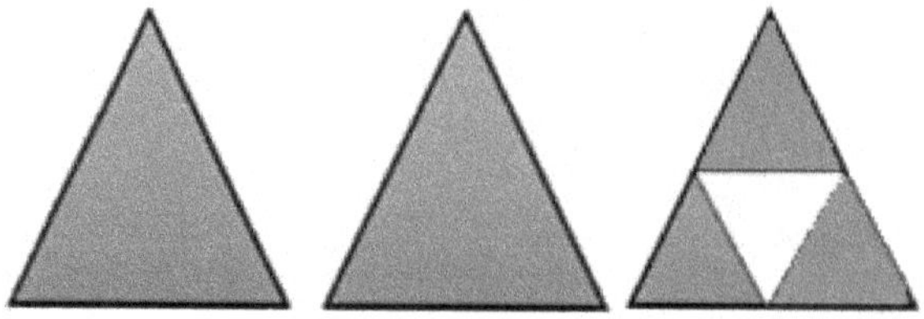

24) A straight-line measure 180°. A straight line and a triangle are touching as shown in the figure below.

What is the value of *A* in the figure?

A. 60

B. 20

C. 90

D. 110

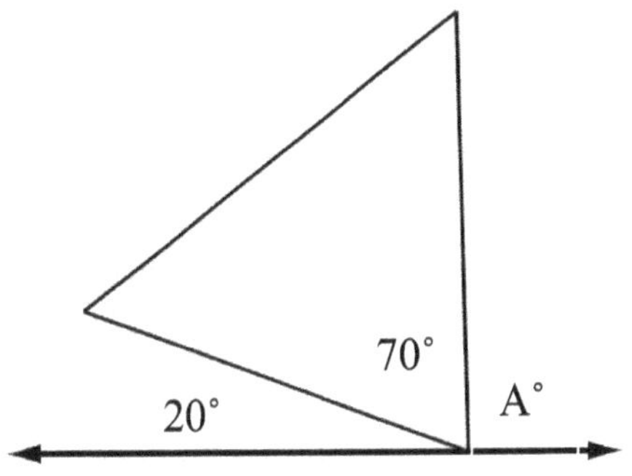

25) The number 64.05 can be expressed as ________

A. $(6 \times 10) + (4 \times 1) + (5 \times 0.01)$

B. $(6 \times 10) + (4 \times 1) + (5 \times 0.1)$

C. $(6 \times 1) + (4 \times 1) + (0 \times 1) + (5 \times 1)$

D. $(6 \times 10) + (4 \times 1) + (0 \times 10) + (5 \times 100)$

26) What is the perimeter of this shape?

A. 72

B. 54

C. 30

D. 33

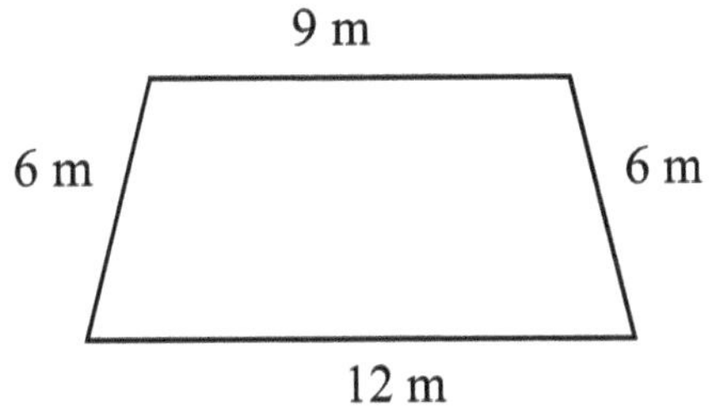

27) The temperature on Sunday at 12:00 PM was 92°F. Low temperature on the same day was 53°F cooler. Which temperature is closest to the low temperature on that day?

A. 58°F

B. 39°F

C. 36°F

D. 48°

28) Which shape shows a line of symmetry?

A.

B.

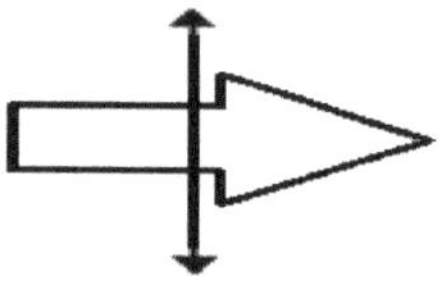

C.

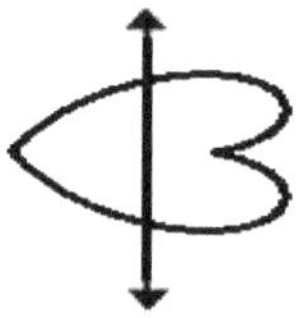

D.

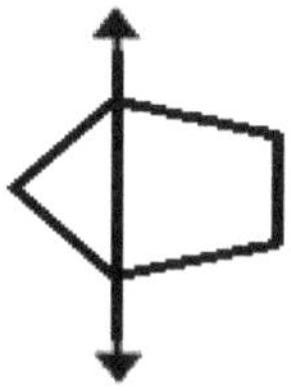

29) There are 365 days in a year, and 24 hours in a day. How many hours are in one-sixth years?

A. 1,340

B. 2,140

C. 1,920

D. 1,460

30) On Sunday Leon was a referee at 4 soccer games. He arrived at the soccer field 20 minutes before the first game. Each game lasted for 20 minutes. There were 10 minutes between each game. Leon left 15 minutes after the last game. How long, in minutes, was Leon at the soccer field?

A. 215 minutes

B. 145 minutes

C. 155 minutes

D. 95 minutes

"This is the end of Practice Test 2."

Chapter 13 : Answers and Explanations

Common Core Practice Tests

Answer Key

⁕ Now, it's time to review your results to see where you went wrong and what areas you need to improve!

Common Core- Mathematics

Practice Test - 1					
1	D	**11**	D	**21**	B
2	B	**12**	B	**22**	A
3	C	**13**	B	**23**	B
4	680	**14**	B	**24**	B
5	A	**15**	A	**25**	C
6	C	**16**	A	**26**	C
7	C	**17**	B	**27**	C
8	C	**18**	A	**28**	C
9	B	**19**	C	**29**	A
10	C	**20**	D	**30**	C

Practice Test - 2					
1	C	**11**	84	**21**	$729\ in^3$
2	D	**12**	C	**22**	B
3	D	**13**	C	**23**	D
4	C	**14**	B	**24**	C
5	A	**15**	B	**25**	A
6	C	**16**	B	**26**	D
7	C	**17**	A	**27**	B
8	C	**18**	A	**28**	A
9	A	**19**	A	**29**	D
10	D	**20**	D	**30**	B

Practice Test 1

Common Core- Mathematics

Answers and Explanations

1) Answer: D.

$50 \times 32 = 1{,}600$.

2) Answer: B.

Tam wants to divide his 350 cards into boxes of 70 cards. Therefore, he needs $350 \div 70 = 5$ boxes.

3) Answer: C.

The clock shows 5:15 in the noon. 2 hours ago, it was 3:15 AM. 45 minutes before that was 2:30 AM.

4) Answer: 680.

We round the number up to the nearest ten if the last digit in the number is 5,6,7,8, or 9

We round the number down to the nearest ten if the last digit in the number is 1,2,3, or 4

If the last digit is 0, then we do not have to do any rounding, because it is already to the ten. Therefore, rounded number of 679 to the nearest ten is 680.

5) Answer: A.

Use area of rectangle formula. Area = length × width

The length of the rectangle is about 4 inches and its width are about 3 inches.

Therefore, the area of the rectangle is $4 \times 3 = 12$ inches.

6) Answer: C.

7 days = 1 week, 31 days = $(31 \div 7) = 4.29 \Longrightarrow 4$ weeks

7) Answer: C.

The football team should buy 15 uniforms and each uniform cost \$19. Therefore, they should pay (15 × \$ 19) \$285.

Choice C is correct answer:

$(19 \times 10) + (19 \times 5) = 190 + 95 = \$285$

8) Answer: C.

20 + Z = 55. Therefore, Z = 55 – 20 = 35

Let's review the options provided.

A. 55 – 20 = Z	Yes! This is true. 55 – 20 = 35
B. 55 – Z = 20	Yes! This is true. 55 – 35 = 20
C. Z – 20 = 55	No! This is not true. 35 - 20 = 15
D. Z + 20 = 55	Yes! This is true. 35 + 20 = 55

Option C is the only option that is NOT true.

9) Answer: B.

Three – digit even numbers that have a 5 in the hundreds place and a 3 in the tens place are 530, 532, 534, 536, 538.

Option B, 532, is the answer.

10) Answer: C.

The number of votes for Football was 9, for basketball was 4, for soccer was 10, and for volleyball was 2. Only table C shows all these numbers correctly.

11) Answer: D.

There are 100 equal parts. 66 parts are shaded. It is equal to $\frac{66}{100}$ or 0.66.

12) Answer: B.

Choice B shows the numbers in order from least to greatest.

21,851< 28,081 < 30,032< 37,009

13) Answer: B.

6.0 – 4.75 = 1.25

14) Answer: B.

Add adult tickets and children's ticket that have been sold.

$$\frac{52}{100}+\frac{1}{10}=\frac{52}{100}+\frac{10}{100}=\frac{52+10}{100}=\frac{62}{100}=\frac{31}{50}$$

15) Answer: A.

This angle is less than 180° and more than 90°. Only choice A.

16) Answer: A.

$\frac{51}{10}$ is equal to 5.1.

17) Answer: B.

Only on option B, all shapes have at least one set of parallel side and one set of perpendicular side.

18) Answer: A.

Add the number of students at the library and number of students in classroom.

$58 + 17 = 75$

19) Answer: C.

For the height of 1 centimeter, we have 4 pennies, therefore, for the height of 16 centimeters, we have 4×16 pennies.

20) Answer: D.

The first model from left is divided into 6 equal parts. 4 out of 6 parts are shaded. The fraction for this model is $\frac{4}{6} = \frac{2}{3}$. The second model is divided into 8 equal parts. 4 out of 8 parts are shaded. Therefore, the fraction of the shaded parts for this model is $\frac{4}{8} = \frac{1}{4}$.

These two models represent different fractions.

21) Answer: B.

$24 \div A = 4$, therefore, $A = 24 \div 4 \rightarrow A = 6$

22) Answer: A.

When rounding to the nearest thousand, you will need to look at the last three digits. If the last three digits is 499 or less round down the number ending with three zeros.

On the other hand, if the last three digits is 500 or more, round up the next ending with three zeros. Since, in the number 7,143, the number 143 is smaller than 499, then round the number to 7,000.

23) Answer: B.

$\frac{75}{100}$ is equal to 0.75.

24) Answer: B.

$12 \div 2 = 6$.

25) Answer: C.

$\frac{8}{1,000}$ is equal to 0.008.

26) Answer: C.

Only option C fits Emma's description.

27) Answer: C.

$12 \times 24 = 288$.

28) Answer: C.

$\frac{7}{17} - \frac{6}{17} = \frac{1}{17}$.

29) Answer: A.

If one box has 80 flashcards, therefore, 50 boxes have the capacity of (50×80) 4,000 flashcards.

30) Answer: C.

0.13 is equal to $\frac{13}{100}$.

Practice Test 2

Common Core- Mathematics

Answers and Explanations

1) Answer: C.

A. The digit 9 has a value of $9 \times 10,000$, not $9 \times 1,000$.

B. The digit 5 has a value of $5 \times 1,000$, not 5×100.

C. The digit 7 has a value of 7×10. This is true!

D. The digit 4 has a value of $4 \times 100,000$, not $4 \times 10,000$.

2) Answer: D.

use perimeter of rectangle formula. P = 2 (length + width)

$P = 2 \times (4 + 8) = 2 \times 12 = 24$ cm

3) Answer: D.

The difference of each two successive numbers is 250.

Add four 250 to last number (2,200):

$2,200 + 250 + 250 + 250 + 250 = 3,200$

4) Answer: C.

4 quarters = 4×25 pennies = 100 pennies

8 dimes = 8×10 pennies = 80 pennies

In total Nicole has 205 pennies

5) Answer: A.

1 issue of the magazine = \$8.

\$96 ÷ \$8 = 12 issues of the magazine.

6) Answer: C.

To find the perimeter of a triangle, add all three sides of the triangle.

$P = 21 + 28 + 35 = 84$ inches

7) Answer: C.

In any square, both diagonals have equal measure and are perpendicular.

8) Answer: C.

To compare fractions, we can write fractions with the same denominator. Then, we can compare the numerators of each fraction and put them in correct order from least to greatest or greatest to least.

Common denominator of 8, 2, 6 and 24 is 24. Rewrite the fractions:

$\frac{5}{8} = \frac{15}{24}$ $\frac{1}{2} = \frac{12}{24}$ $\frac{5}{6} = \frac{20}{24}$ $\frac{1}{24} = \frac{1}{24}$

9) Answer: A.

$52 + 9 = 61.$

10) Answer: D.

\$31+ \$14 = \$45.

11) Answer: 84.

1 bag = 15 red cards + 6 black cards (15 + 6 = 21 cards)

4 bags = 4 × 21= 84 cards.

12) Answer: C.

To round numbers to the nearest ten thousand, make the numbers whose last five digits are 00001 through 49999 into the next lower number that ends in 00000. For example, 724,424 rounded to the nearest one hundred thousand would be 700,000. Choice C is correct because last five digits of 54,723,330 is less than 49,999.

13) Answer: C.

1 year = 48 games. 5 years = 5 × 48 = 240 games

14) Answer: B.

Use perimeter of rectangle formula. P = 2 × (Length + Width)

26= 2 × (6 + W) ⇒ 26 = 12 + 2 × W ⇒ 2 × W = 26 – 12 = 14 ⇒ W = 7 feet.

15) Answer: B.

3 feet = 1 yard;   63 feet = 21 yards.

16) Answer: B.

The sum of A and B equals 35: A + B = 35, if A = 18,

then B = 35 – 18 = 17.   18 + 17 = 35 or B + 18 = 35

17) Answer: A.

A = 172 ÷ 4 = 43

18) Answer: A.

To compare fractions, we can write fractions with the same denominator. Then, we can compare the numerators of each fraction and put them in correct order from least to greatest or greatest to least.

Common denominator of 2, 4, 8, and 32 is 32. Rewrite the fractions:

A. $\frac{1}{32}$ B. $\frac{1}{2}=\frac{16}{32}$ C. $\frac{3}{4}=\frac{24}{32}$ B. $\frac{5}{8}=\frac{20}{32}$

19) Answer: A.

An obtuse triangle is one with one obtuse angle (greater than 90°) and two acute angles. Since a triangle's angles must add up to 180°, no triangle can have more than one obtuse angle.

Only shape A has one obtuse angle.

20) Answer: D.

16 pastilles = 1 box

64 pastilles = (64 ÷ 16) = 4 boxes

21) Answer: 729.

To find the volume of cube, multiply one side of the cube by itself 3 times:

Volume of a cube = (side) × (side) × (side) = 9 × 9 × 9 = 729 in^3.

22) Answer: B.

Arrow is pointing to a number in the middle of two numbers 10 and 18. Therefore, the answer is 14.

23) Answer: D.

This shape shows 2 complete shaded triangles and a three of a four part of triangle.

The mixed number for this shape is $2\frac{3}{4}$.

24) Answer: C.

Three angles in a triangle add up to 180°.

A° + 70° + 20° = 180° ⇒ A° = 180° – 90° = 90°

25) Answer: A.

The number 64.05 can be expressed as:

$(6 \times 10) + (4 \times 1) + (5 \times 0.01) = 60 + 4 + 0.05 = 64.05$

26) Answer: D.

To find the perimeter of the shape, add all four sides.

$P = 6 + 12 + 6 + 9 = 33$

27) Answer: B.

Low temperature is 53°f cooler than the temperature at 12:00 PM which is 92°. Low temperature is 39°f (92°f – 53°f) that is choice B.

28) Answer: A.

You can find if a shape has a Line of Symmetry by folding it. When the folded part sits perfectly on top (all edges matching), then the fold line is a Line of Symmetry. Shape A shows a line of symmetry.

29) Answer: D.

1 year = 365 days, 1 day = 24 hours

$\frac{1}{6}$ years $= \frac{1}{6} \times 365 \times 24 = 1{,}460$ hours

30) Answer: B.

Each game 20 minutes, therefore 4 games took $4 \times 20 = 80$ minutes.

10 minutes between each game. There are 30 minutes in total between 4 games. (between game 1 and 2, 10 minutes and between game 2 and 3, 10 minutes and between game 3 and 4, 10 minutes)

Leon arrives 20 minutes before first game and left 15 minutes after the last game.

In total, he was $80 + 30 + 20 + 15 = 145$ minutes at the soccer field.

"End"

www.ingramcontent.com/pod-product-compliance
Lightning Source LLC
Chambersburg PA
CBHW080442110426
42743CB00016B/3247
* 9 7 8 1 6 3 6 2 0 0 9 2 7 *